The American Royal: 1899-1999

The American Royal: 1899-1999

by

Heather N. Paxton

Underwriting provided by
**Enid and Crosby Kemper Foundation
UMB Bank, n.a.**

A Wallaroo Book
BkMk Press, University of Missouri-Kansas City

in cooperation with
The American Royal Association

Cover design by Dennis Forbis

A Wallaroo Book   American Royal Association
BkMk Press, University of Missouri-Kansas City   1701 American Royal Court
5100 Rockhill Road   Kansas City, Missouri 64102
Kansas City, Missouri 64110

Library of Congress Cataloging-in-Publication Data

Paxton, Heather N.
        The American Royal: 1899-1999 / Heather N. Paxton.
                p.   m.
        Includes bibliographical references (p.   ).
        ISBN 1-886157-22-7 (paper). – ISBN 1-886157-23-5 (cloth)
        1.   American Royal (Livestock show)—History. I. Title.
        SF117.65.M82K366   1999
        636.08'11—dc21   99-32001
                                                                CIP

        Printed in the United States of America
        by Walsworth Publishing Company, Marceline, Missouri

        10 9 8 7 6 5 4 3 2 1

Photos used in the cover collage also appear inside the book; please refer to captions within the text for photo credits.

Contents

Author's Note   vii

Preface   xi

Prologue   xiii

The Early Years   3

The 1920s   17

The 1930s   30

The 1940s   44

The 1950s   49

The 1960s   61

The 1970s   69

The 1980s   79

The 1990s   85

1998   93

The Centennial   95

The Future   97

Appendices   99

Bibliography   127

Author's Note

First of all, I would like to acknowledge R. Crosby Kemper, Jr., without whom this book, and so many other worthwhile endeavors in Kansas City, would not have been possible.

When I began writing this book, Landon Rowland was president of the American Royal. He was succeeded by Fred W. Lyons, Jr., the current president. I wish to thank both gentlemen for their support of this project.

A book such as this one is not the work of one person. I owe a great debt to the wise counsel of the history committee: John A. Dillingham, chairman, Mary Hunkeler, James W. "Jim" Leathers, Jean Riffle, Col. John A. Riffle, USAF (Ret.), Melissa "Missy" Shores and B. C. "Bud" Snidow.

I was very fortunate to have the able assistance of James D. "Jim" Taylor, vice president and general manager of the American Royal, Robert D. "Bob" Hovey, legal counsel to the American Royal, and Colleen Scott, who served as the American Royal's liaison to the history committee. I also would like to thank Phyllis Barragan, Fern Palmer Bittner, Lisa Cabra, Bev Chester, Marj Collins, Dolores Ford, Scott Meggs, Yvonne Miller, Teresa Morgan, Yvonne Miller, Luanne Neuner, Paulette Orth, Nancy Perry, George Rush, Cindy Stanley, Jenny Stuerzl, Penny Yates and Marilyn Zerbe, as well as Donna J. Brady, Jack Clark, Flo Coffey, Evelyn and George Crews, Liz Goede, Connee Hays, Bobby Holliday, Bill Kimball, Stacy Maher, Pene Maurer, Lindsay McLean, Marjorie Norman, Trcy Satterfield, Beverly Schroeder, Doris and Bud Sloan, Mike Urbom and Sandi Winkfield, all of the American Royal.

Professor James McKinley of the University of Missouri-Kansas City has been involved with this project for nearly a decade. I am grateful to him for his fine work as an editor and as my teacher, as well as for the invaluable oral interviews and questionnaires he collected in the early 1990s. Professor McKinley was assisted by Glenda McCrary, who deserves praise for transcribing the interview tapes, thus simplifying my work considerably. Those who took the time to share their memories with Professor McKinley and Ms. McCrary are Ray Davis, Jay B. Dillingham, Frances Dillingham, Jo Ann Field, William Harsh, Neville Hunsicker, Donald Ornduff, Laurence Pressly, George Shepherd, Art Simmons, Joey Straube, Rosemond "Roz" Straube and Edwin Wade "Eddie" Williams.

I deeply admire and appreciate the foresight of Joanne Faulkner in tape recording the memories of four long-time American Royal participants, Jay B. Dillingham, Ralph Peak, Art Simmons and

Harold Thieman in the early 1980s. Mrs. Faulkner was one of many Royal enthusiasts who wrote out their recollections for a questionnaire in the early 1990s. The book is richer for her contributions and those of these ladies and gentlemen: Mary Loy Brown, Redd Crabtree, Cecil L. Eyestone, Don L. Good, Carol Collins Hovey, Robert D. Hovey, Charles W. "Bud" Keller, Karen Greer Lesmeister, James S. Lowrey, Barbara Mainord, John E. "Johnie" Miller, Sue Noll, Audrey R. Peak, W. C. Theis, Harold W. Thieman, Connie L. Warren, Lynn Weatherman and E. W. "Eddie" Williams.

David Boutros, Marilyn Burlingame, Jennifer Parker and Betty Swantig at the Western Historical Manuscripts Collection at the University of Missouri — Kansas City were tireless in helping me locate key materials in boxes of archival documents.

For this book, I interviewed more than thirty people, each of whom selflessly spent hours with me and all of whom were unfailingly patient with me in explaining details and reviewing events. I am most grateful to: Anthony P. Arnold, Malcolm M. "Mick" Aslin, Fern Palmer Bittner, Frances Dillingham, Jay B. Dillingham, John A. Dillingham, Joan Edwards, Cecil L. Eyestone, John C. Gage, Dr. Don L. Good, Amy Haun, R. E. "Bud" Hertzog, D.V.M., Carol Hovey, Robert D. "Bob" Hovey, Mary Hunkeler, Charles W. "Bud" Keller, R. Crosby Kemper, Jr., Olive Lansburgh, James W. "Jim" Leathers, Gladys Mackey, Paulette Orth, Shirley Parkinson, Nancy Perry, Col. John A. Riffle, USAF (Ret.), Marie Schubert, Melissa "Missy" Shores, Roger Shores, Kathryn "Kit" Smith, Robert E. "Dr. Bob" Smith, D.D.S., B. C. "Bud" Snidow, Cindy Stanley, James D. "Jim" Taylor, Sug Utz, Betty Weldon and F. George Zahn. Undoubtedly, Jay B. Dillingham has been the most generous with his time: He has been interviewed through the years by Mrs. Faulkner, Professor McKinley and me, as well as by the *Kansas City Star* and other publications.

Amy Haun, co-chairman for the book, *Belles of the American Royal Fiftieth Anniversary 1949-1999*, spent many hours discussing the BOTAR Organization and the Royal with me and helping me locate files and photos. Georganne Hall of *The Independent* magazine received several calls from me, and with the assistance of her mother, Sallie Oliver (Mrs. William W. Oliver), provided me with missing names. Wilma Hughes of *The National Horseman* not only answered numerous questions about the horse show, she also arranged the loan of many photographs for the book. All of these ladies have gone out of their way to help me, and I deeply appreciate their aid.

Ben Furnish and Roxanne Witt of BkMk Press have painstakingly seen this project through its developmental processes, overseeing the completion of the text, the addition of photographs and, ultimately, this book as it now appears.

Other organizations and individuals who assisted me in my research include: the American Hereford Association: Janell Coe and Craig Huffines; Larry Atzenweiler; the Black Archives of Mid-America Inc.: Bill Livingston; Jim Blair; the BOTAR Organization; Michael S. B. Churchman; the Cowboy Hall of Fame; Alberta Lee Cox; Culver Military Academy; Tom Devine; Stephanie Doerflinger; the Future Farmers of America: Coleman Harris, Kevin Keith and Laura Thomas; Eleanor Gage; Brandon Goehring; Janet Green; Joyce Hale; Margaret Hall; Susan Hand; Hulse Stables: Nancy Cummins and Don Hulse; *The Independent* magazine: Laureen Ingram; the Jackson County Historical Society: Cora J. Wilson; the Johnson County Museum: Kathy Stump; Linda Newcom Jones; the *Kansas City Star*; the Kansas State Library; Kansas State University: Fred Delano and Daisy Fielder; Mary Keaveny; Alexander C. "Sandy" Kemper; Judy Kendrick; Betty and Bucky Kessinger; Marianne Kilroy; Bob Kraut; Vicky Leonard; Tod Macklin; James Madison, first national vice president, 9th and 10th Horse Cavalry Association; Larry E. Mead; the Missouri Department of Agriculture: John Saunders; Missouri Valley Special Collections, Kansas City Public Library; George Morse; the National Association of State Universities and Land-Grant Colleges: Barbara Cummings; Dana Hale Nelson; Mary Nofftz; Isabel and Frank Paxton III; Margie Paxton and Preston McCall; Millie Paxton; Commander Carleton Philpot, US Navy (Ret.); Plaza III: Joe Wilcox; Laurence Pressly; Hayley Walker Rees; the

Will Rogers Memorial Museum: Greg Malak; Wanda and John Royle; the Saddle & Sirloin Club: Donna Snelson; H. Leon Sargent; the Savoy Grill; Steve Schneider; Betty and Tom Scott; Dr. Kimberly Shara; Joey Straube; Loretta Terbovich; George Terbovich; the Harry S. Truman Library: Dennis Bilger, Liz Safly, Anita Smith and Pauline Testerman; UMB Bank, n.a.: Sarah Cox and Tonya Whetsel; the University of Missouri — Columbia: Dave Baker; Claudette Walker; Anne Turner Wells; Walker Woods and Sue Zahn.

Unfortunately, some historical documents have been lost, (especially as a result of the floods of 1993 and 1998). In addition, many of the photographs from the archives are lacking complete identifications. After much consideration, the history committee and I chose to publish the ones we felt best represented the American Royal. If you have any information regarding the photographs in this book or other American Royal activities and memorabilia that you would like to share, please write to:

> The American Royal
> 1701 American Royal Court
> Kansas City, MO 64108
> Attn: Museum and Visitors Center

The telephone number at the Royal is (816) 221-9800.

As is apparent, I was the beneficiary of a tremendous amount of information and material from a multitude of sources, all of which I deeply appreciate. I have done my level best to review, compile and incorporate that material into this book. If I have overlooked anybody, any photo or any document, or the mention of any significant contributor, to our beloved American Royal, I tender my regrets. It has been a wonderful experience for me and I hope you will enjoy the results of my efforts to chronicle the one hundred years of history of the Royal.

The proceeds from the sale of this book will benefit the American Royal Metropolitan Scholarship Fund, an endowed educational fund for Kansas City area youth.

> H. N. P.
> Kansas City, Missouri
> July 8, 1999

Preface

American Royal Centennial.................100 Years!

Much has happened in the period of 1899-1999. The changes the world has experienced through the industrial revolution, World Wars I and II, major breakthroughs in technology — from the automobile and airplane to computers and biotechnology medications — are almost beyond comprehension.

The world of agriculture and rural America has changed dramatically during this period as well. As recently as the 1940s, agriculture labor made up 40% of the labor force. With technological advancements, productivity and quality of agricultural products has increased while the percentage of the total labor force in agriculture has dropped to 2 - 3%.

Rural America, built upon the foundations of integrity, family values and hard work, has been impacted. While it struggles, it is beginning to succeed in applying those values and skills to new endeavors and new productive enterprises. Over these years the American Royal has continued to expand and meet the needs of the agriculture and rural America constituencies.

This book enlightens us as to how the Royal expanded from a group of cattlemen gathering to exhibit their Hereford cattle in 1899 to the largest livestock, horse show and rodeo in America. Our gratitude is extended to the leaders, staff and many volunteers who have made this progress possible.

The educational focus of the Royal and the "fun" focus of its activities give a strong direction of its future opportunity and challenges as we approach the new millennium. We hope you will continue to be an active participant in this celebration of such a vital area of American culture and economic progress.

Special thanks are extended to the members of the history committee that worked so diligently to bring to us this insight into our past:

> John Dillingham, chairman
> Mary Hunkeler
> James W. "Jim" Leathers
> Colonel and Mrs. John Riffle
> Melissa "Missy" Shores
> B. C. "Bud" Snidow

We also acknowledge Landon Rowland, president of the American Royal at the time these efforts were initiated, for his foresight and vision to provide this document.

See you at the Royal!

Fred W. Lyons, Jr.
President
American Royal

Prologue

The thrill of the American Royal: You can see it in the children's eyes. Some stayed up late to come to the Wild West Show and see their first horses. Others awakened early, knowing that this was the day they would show and sell the hogs they had raised. Others stood nervously in the stalls, brushing their horses as they waited to ride around the ring. Some came with school groups, shrieking with delight while patting the goats in the petting zoo. These memories will stay with them. And they'll remember the cattle: the red heifer who flicked her tail as they walked by, the black steer who pawed the ground with one heavy hoof and snorted, creatures more enormous than they had ever imagined — and some even had horns, great big ones.

Upon leaving Kemper Arena after the rodeo, the children carefully circle the "Bull Wall" sculpture, looking through it to see the American Royal Center and imagining the past. *What was Kansas City like a hundred years ago?* they wonder silently, as they shade their eyes against the autumn sun . . .

"Bull Wall" (Photo courtesy of the American Royal Association).

The Early Years

October, 1899: Foreign correspondents for American newspapers were filing dispatches from the Philippines. Guglielmo Marchese Marconi was experimenting with wireless telegraphy between battleships (he would win the Nobel Prize for physics in 1909). Helen Keller was studying mathematics, languages, history and natural science at Radcliffe College. Alexander Kissee, a 70-year-old Missouri man known as the "king of Taney County" was divorced from 15-year-old Dora Garrett after a marriage that lasted a week; the bride resumed use of her maiden name and was granted $4,000 in alimony.

In Kansas City, a realtor's ad headlined "What Do You Want? You Can't Beat This" offered a 10-room brick house on 17th and Wabash streets for $2,600. The Palace Clothing Co. was selling "men's stylish novelty worsted suits" for $9.50. Emery, Bird, Thayer featured "women's automobile coats" from $19.75 to $55. The *Kansas City Star* announced that strawberries from Florida were expected in stores soon; California celery was already on the shelves. Madame Calve, the opera singer, was coming to town. The Horse Show was underway at Convention Hall. Tom Bass' saddle mare, Miss Rex, so enchanted the crowd by waltzing to music that one woman threw her a bouquet of American Beauty roses. And something new was starting: the American Royal.

The American Royal started in 1899 as the National Hereford Show, the first nationwide show for the exposition and sale of purebred cattle. The show's organizers included C. R. Thomas, Charles Gudgell, Thomas Clark, H. H. Clough, F. A. Nave, John Sparks, James A. Funkhouser, Tom Smith, C. A. Stannard and T. F. B. Sotham. It was held in a tent in the West Bottoms from October 23-28, 1899 and featured 541 registered Herefords. During the three-day sale, 300 Herefords were sold, at an average price of $334. The estimated attendance was 55,000.

From the very beginning, the show was a success. On October 28, 1899, the *Kansas City Star* recorded the favorable opinion of visitors to the National Hereford Show held in Kansas City:

"This is the place for a Hereford show," said John Hooker of New London, O. "I never was so pleased with any town I ever was in before. The big breeding farms and ranches where beef cattle are grown center around here. It is just the place for Hereford men to come, the place for big shows. The only thing I regret is that I don't own a breeding farm within a day's ride of this town . . . I don't know where else a Hereford show could

3

These pictures depict the 1899 National Hereford Show — the first American Royal (Photos courtesy of the Western Missouri Historical Collection-Kansas City, left, and the American Royal Association, right).

be held to such advantage."

Hereford cattle had gained popularity in the Midwest in a relatively short period of time. Charles and James Gudgell and Thomas A. Simpson had been Shorthorn breeders in Pleasant Hill, Missouri prior to 1876, when they attended the Centennial Exposition in Philadelphia and were impressed by the Hereford display. The firm, which later moved to Independence, Missouri, soon began purchasing Herefords. In May 1879, the firm sold several bulls at an auction in Kansas City; according to Charles Gudgell, who was president of the American Hereford Cattle Breeders' Association during the first seven years of the American Royal, the 1879 sale was the first public auction of purebred Herefords west of Ohio. Gudgell & Simpson imported Anxiety 4th, a bull from Herefordshire, England in 1881. Thomas A. Simpson believed that the young bull was ideal for breeding purposes: He was right. Anxiety 4th 9904 would become known as "the father of American Herefords." His progeny would include Don Carlos, Beau Brummel, Lamplighter, Prince Domino and Beau Mischief. As Donald R. Ornduff wrote in *The First 49*, "virtually all American Herefords 100 years after the birth of Anxiety 4th are descended from him."

From its earliest days, the show was characterized by civic pride and a spirit of generosity. The most poignant event of the 1899 show was a shower of silver dollars. Kate Cross, the widow of Hereford breeder Charles Cross, had brought the last of his herd, a young bull named Bonnie Prince, to be sold. Jno. M. Hazelton, historian and editor of the *American Hereford Journal*, recalled the events for a 1933 anniversary edition of the journal: While the bidding on Bonnie Prince was in progress, Col. C. C. Slaughter, Dallas, Texas, who was at ringside, said he regretted that he could not bid on the bull because he was too closely related to the cows in his herd, but he wanted to see Mrs. Cross realize as much as possible from his sale, and tossed a silver dollar into the ring. For a few minutes silver dollars rained into the ring from every section of the sale pavilion. Mrs. Cross received $1,100 for the bull in addition to the silver dollars. She later established her own herd of Herefords.

Kirkland B. Armour, the American Royal's first president and a member of the Armour family, which owned a meat-packing business, offered a heifer named Armour Rose as the prize in a lottery to benefit the building fund for the proposed Convention Hall. Mr. Armour next bought the heifer back from the lady who won the raffle. He then put Armour Rose up for auction at the cattle sale at the National Hereford Show. Armour Rose sold to John Sparks, former governor of Nevada, for $2,500, which Mr. Armour contributed to the building fund. The Nebraska Clothing Co. celebrated the event in an advertisement in the *Kansas City Star*:

**Anxiety 4th 9904 (above, sketch by N. A. Throop);
Kate Wilder Cross (right)** (Both from *The Story of the
Herefords* by Alvin H. Sanders).

Armour Rose to the occasion yesterday, and showed his good judgment by getting the highest price ever paid for any "So boss" in the world. Marshall Field wanted her, but she didn't want to go to Chicago Fields — she wanted to go West to other fields, and her wish will be gratified, for along comes John and Sparks her and wins her.

The Convention Hall was built and Kansas Citians were eagerly anticipating the July 4th opening of the Democratic National Convention when the building burned on April 4, 1900. The hall was completely rebuilt in three months — in time to be the site of the nomination in absentia of William Jennings Bryan. The American Royal Horse Show was held in Convention Hall in 1915 and 1916.

Mr. Armour also gave the Armour Cup, valued at $400, which was awarded to Dale 66481, the Grand Champion bull, who was owned by Frank A. Nave of Attica, Indiana. Fat cattle were highly prized in that era. According to Alvin Sanders in his 1914 book, *The Story of the Herefords*: "Dale was bred by Clem Graves of Bunker Hill, Ind., being sired by Columbus 51875 and out of Rose Blossom . . . Dale had put on flesh about as thickly as a compactly fashioned bovine carcass ever carries, and shared with the heifer Armour Rose the adoration of the Hereford-worshipping multitudes that thronged this sensational ringside during the most memorable week of American Hereford history up to that date."

Dale's son, Perfection, shown by Thomas Clark of Beecher, Illinois, won the Armour Cup in 1900. Dandy Rex, from the show herd of Gudgell & Simpson in Independence, Missouri, was the 1901 winner. J. A. Funkhouser of Plattsburg, Missouri, owned the 1902 winner, March On 6th, and the 1903 winner, Onward 4th. The Armour Cup itself disappeared for many years. It once sold at a garage sale for $2.50. In 1981, Bud Snidow of the American Hereford Association was contacted by its then owner, who wished to sell it. The Harry Darby Foundation purchased the cup, which is on display at the American Hereford Association headquarters. A replica of the cup has been presented as the Harry Darby Award to the owner of the grand champion Hereford bull in recent years; the award will be retired after the centennial.

Kirkland B. Armour began serving as president of the show in 1899; he was also president of the American Hereford Association at that time. He is listed as president of the American Royal through 1902, but he died in September, 1901 at the age of 47. His farmhouse at 67th Street and Pennsylvania is still standing. The Armour Hills and Armour Fields neighborhoods are named for him.

In addition to Mr. Armour, Charles R. Thomas and Eugene Rust also are contenders for the

Thomas A. Simpson (left) and Charles Gudgell (right). (Photos from *The Story of the Herefords* by Alvin H. Sanders).

Kirkland B. Armour (left) and Frank A. Nave (right). (Photos from *The Story of the Herefords* by Alvin H. Sanders).

title "Father of the American Royal." As secretary of the American Hereford Association, Mr. Thomas was instrumental in staging the National Hereford Show. "It was his brainchild, really, that started the Royal," said Bud Snidow. "He went to the Columbian Exposition in Omaha and said 'Why can't we do something this big, strictly for Herefords?'" Mr. Thomas was the general manager of the show in 1904 and 1905 and served as American Royal president in 1908. Mr. Rust, who was superintendent of the Kansas City Stock Yards Company in 1899, served as general manager of the Royal in 1903 and as president in 1907.

It should come as no surprise that a livestock show would be a success in Kansas City, which has often been considered "the Gateway to the West."

Osage Indians had been the original settlers of this land, but by the early 1800s the United States government was starting to take an interest in the area. Despite Army officer Stephen H. Long's 1819 report that the lands west of the Missouri River from Kansas into Colorado were "almost wholly unfit for cultivation and, of course, uninhabitable by a people depending upon agriculture for their subsistence," both the town of Kansas (now downtown Kansas City) and the town of Westport began to grow in the 1830s. The California Gold Rush spurred travelers into the town: 40,000 people passed through in 1850. That year, the city's 700 inhabitants decided to incorporate as Kansas City. In 1856, two years after Kansas became a state, Alexander Majors, of the freighting firm Russell, Majors & Reed, built a house on the Missouri side of the state line: From his window, he could see his cattle grazing in Kansas. This house, south of the city in those days, is still standing at 82nd Street and State Line Road.

The Battle of Westport in 1863 scarred the area south of 40th street: In the early 1900s, golfers on the course of the origi-

The Armour Cup (By permission of the American Hereford Association, courtesy of the Western Missouri Historical Manuscript Collection-Kansas City).

Dale 66481 (Drawing from *The Story of the Herefords* by Alvin H. Sanders).

7

Charles R. Thomas (From *The Story of the Herefords* by Alvin H. Sanders).

nal Kansas City Country Club (now the site of Loose Park) occasionally dug up a minie ball. Kansas City might have lost momentum with the war, but two events help assure the continued growth of the Midwest: the opening of land-grant schools and the brief era of the cattle drives.

The Morrill Act, which was passed by Congress in 1862, gave each state 30,000 acres of land for each representative the state had in Congress. Ninety percent of the proceeds of this land were to be used to set up and maintain land-grant colleges and universities. These schools would teach agriculture and mechanical arts. "The object of the land-grant institutions was to give youths the opportunity to have an education, and especially an agricultural education," according to Dr. Don Good, retired head of the department of animal science at the Kansas State University. Agricultural experiment stations at the schools conducted research to improve the development of crops and livestock. The three basic goals of the land-grant schools were to provide academic education, to foster research opportunities and to develop extension services. The 4-H Clubs were created to train farm youths in agriculture and are an extension service of the land-grant schools. Universities in the Midwest that were founded as land-grant schools or which received additional funding as a result of the Morrill Act of 1862 include the University of Missouri, Kansas State University, Oklahoma State University, Iowa State University, Colorado State University, Michigan State University and Ohio State University. The Second Morrill Act, which was passed by Congress in 1890, provided for the creation of land-grant schools for African-Americans.

It is remarkable to realize that Kansas State University was founded as a land-grant school in Manhattan, Kansas in 1863 — during the Civil War. Other projects had been postponed and sabotaged by the coming of the war. The Pacific Railroad of Missouri had begun work on a railroad to run from St. Louis to Kansas City in 1851. Railroad workers were still laying track during wartime, but the railroad was a target of Confederate soldiers, who tore up tracks and blew up bridges. After General Sterling Price's 1864 raid, in which bridges over the Gasconade and Osage Rivers were destroyed, some feared the railroad would never be completed. Instead, crews worked steadily for five months, from the end of the war in April, 1865 until that September, when the line was completed. The first train pulled into Kansas City's east bottoms on September 25, 1865 — six days after the last spike had been hammered into place.

Less than four years later, on July 3, 1869, the Hannibal Bridge opened. The bridge, which was designed by engineer Octave Chanute (who used as his model a bridge over the Rhine), made it possible for trains to cross the Missouri River. Kansas City quickly became a major artery for the railroads: Forty years later, 39 different rail lines would pass through the city. Kansas City was already experiencing growth. It had been a town of 3,500 inhabitants at the end of the war; by 1870, it had a population of 32,000.

In 1867, the Chisholm Trail opened. The trail began in San Antonio, Texas, went north through Indian Territory and ended in Abilene, Kansas, covering a distance of roughly 800 miles. The first year, 35,000 cattle made the trip. Most of the cattle coming from Texas were Longhorns, a breed brought to this continent by Spanish explorers. From Abilene, the cattle could be shipped east on the

4-H Club Conference, 1935 (Photo by Anderson, courtesy of Col. and Mrs. John Riffle).

Kansas Pacific Railroad. The logical marketing center for these cattle? "Kansas City is pre-eminently the point on the Missouri River at which a livestock mart ought to be established and by the united exertions of the western stockmen sustained," wrote cattleman Joseph G. McCoy, who first decided to build the cattle pens by the tiny Abilene station and then to build the town around it.

The peak year for the Chisholm Trail was 1871: An estimated 600,000 cattle journeyed to Abilene. The first stockyards company in Kansas City was founded that year: "It constructed pens, chutes, and other facilities available to all shippers on 13-1/2 acres in the West Bottoms, and added a primitive exchange building. The receipts that year showed a total of 120,000 head of cattle," according to Henry C. Haskell, Jr. and Richard B. Fowler in the book *City of the Future*. Four packing houses were in existence in Kansas City by 1871. The fifth, Plankinton & Armour, opened that year. The Armour family would contribute greatly to Kansas City's history. Kirkland B. Armour, the first president of the American Royal, arrived here in 1870 as a 17-year-old boy. Armour & Co. would become a prominent packing house as would the Cudahy Packing Company, Swift and Company, Morris & Co., Morrell Packing Company and, after World War I, Wilson Packing Company. Commission firms, such as John Clay & Company, Maxwell & Furnish, Martin Bloomquist & Lee, Producers, Swift & Henry, Drumm Standish, & National, served as the middlemen between cattle sellers and the packing plants.

The cattle drives would become a greatly romanticized part of America's memory. Cowboys and chuckwagons still roam the open range on television and in the movies, but the invention of

The Kansas City Stockyards (Photo by James W. Swetnam, used with permission).

barbed wire in 1873 allowed homesteaders to make their lands off-limits to the cattle drovers. By the end of the 1880s, the Chisholm Trail would no longer be in use.

Baylis John Fletcher wrote a book entitled *Up the Trail in '79* that describes how the expansion of the railroads forever changed the frontier. Fletcher was one of a group of cowboys who left Victoria, Texas on March 10, 1879 to drive a herd of cattle up the Chisholm Trail into Kansas and then to Cheyenne, Wyoming. The trip took five months. Along the way, his party endured heavy rains, hail storms, cattle stampedes, quicksand, a scarcity of drinking water and fevers. Rattlesnakes and wolves were occasionally seen and, more often, heard. The drovers did not fight with the Indians they met along the way. Instead, their major arguments were with homesteaders, who feared that the cattle carried disease and also that the cattle would eat crops and destroy farmland. On August 15, 1879, they delivered the herd to purchasers. When Mr. Fletcher decided to return to Texas, the train ride took only a few days. The 1870s would bring another significant change to the Midwest. Mennonite farmers from Russia brought "turkey red" winter wheat to the Kansas farmlands. Kansas City wheat receipts would rise from 678,000 bushels in 1871 to surpass nine million bushels in 1878.

In June, 1876, the Board of Trade began "grain call" trading, the first futures trading in Kansas City. Although a board of trade was first established in 1856 by a group of local merchants and was reorganized as a grain market in 1869, the Kansas City Board of Trade marked its centennial in June, 1976. The Midwest quickly became "the breadbasket of America." As Roger T. Johnson has written:

By 1950, Kansas City had become the second largest milling center in the country, processing grain into flour, syrup, starch, feed, and industrial products. Storage facilities exceeded a capacity of 150 million bushels. The Kansas City Board of Trade had become a center for the agricultural bounty of the Midwest — corn, oats, barley, rye, sorghum, and soybeans — as well as the major world market for hard winter wheat.

Technological advances have changed the ways crops are grown, harvested, packaged and distributed, but the Midwest is still a leading food producer. Or, as a sign on Interstate 70 states, "Every Kansas Farmer Feeds 101 People — And You."

Much would happen in Kansas City in the 1880s and 1890s. William Rockhill Nelson began publishing the *Kansas City Star and Times* in 1880. Mr. Nelson raised Shorthorns on his country property, Sni-A-Bar Farm, in Lee's Summit, Missouri. Mr. Nelson and his neighbor Augustus Meyer, the first president of the Kansas City Parks Department were leaders of the City Beautiful movement which advocated a system of parks and boulevards based on plans drawn by landscape architect George E. Kessler. The Paseo, Cliff

Scenes from the 1900 National Hereford Show— the second American Royal (From *The First 49* by Donald R. Ornduff, with permission of the American Hereford Association).

Drive, Penn Valley Park and Swope Park were created as a result of Mr. Kessler's ideas.

At the turn of the century, Kansas City was a city of twenty-five square miles, with a population of 285,000. The town of Westport had been annexed to Kansas City in 1898.

The 1900 show proved that the success of the 1899 show was no fluke, as recorded in the *Breeder's Gazette* on October 24:

The people came as though summoned to a feast. Millionaire and "cow-puncher", rancher and breeder, farmer and stock-feeder from far and near assembled. The city lent its hundreds of spectators, indicative of the vital interest felt in the industry on which its growth was founded and its prosperity predicted. The vast tent with its capacity for thousands, was repeatedly taxed to the utmost, and officers were unable to keep the passage way cleared to the seats, so determined were the spectators to hug the fence and get as near as possible to the cattle. Idle curiosity played no part in drawing the crowds; interest was the load-stone. And the women were there. They came from the country by the hundreds. No more significant feature of the exhibition can be noted. It indicated the ability of the stock farmer to visit the city with his wife and children; it revealed the interest of the women of the farm in live stock.

The Early Years

The American Royal was given its name after Dean C. F. Curtiss, dean of agriculture at the Iowa State College of Agriculture compared the show favorably with the British Royal Agricultural Fair. The *Daily Drovers Telegram* agreed, proclaiming on January 1, 1901, "Call it the American Royal." That year, Galloway and Shorthorn cattle were shown as well as the Herefords. In 1902, the Aberdeen-Angus Association joined the show. On October 19, 1902, the Kansas City Star reported:

> The crowds of people who swarmed around the American Royal Livestock tents and pavilions yesterday interfered at times with the work of the people who were handling the stock. This amounted to so much that a strict order was issued last night that nobody can be admitted without an exhibitor's, member's or reporter's badge.

Attendance for the show was estimated at 60,000. The badge system proved successful, and is still in use today.

Two events in 1903 set precedents for the American Royal. One was a show of cooperation. In a meeting in January, four men, each representing a different breed of cattle, were selected to manage the American Royal for the year: C. R. Thomas (Herefords), C. E. Leonard (Shorthorns), George Stevenson, Jr. (Angus) and R.W. Park (Galloways). The other was a natural disaster. The history of the American Royal has been marked by floods, the first of which occurred on June 1, 1903. Estimates of the depth of the water in the West Bottoms ranged from 15 feet to 30 feet. The October show went on as before, with the addition of horses and sheep. Among the spectators at the 1903 show were Robert D. and Henry L. Mousel, who were in their mid-20s and already the founders of a famed Hereford-breeding firm in Cambridge, Nebraska. In the years to come, Mousel Bros. would be award-winning exhibitors.

The Missouri mule made its first appearance at the 1904 Royal in the mule show. Missouri mules were big business for many years. Prominent exhibitors included the firm of Herrington & Guyton of Lathrop, Missouri, which exported mules during World War I, and Ferd Owen of Belton, Missouri.

A Galloway calf named Kansas City Royal was born in the show barns on October 17, 1904. Kansas City Royal made his debut in the show ring with his mother, Paragon, the following day. Allen M. Thompson became superintendent of the American Royal that year. Mr. Thompson raised Galloways at his home, Maple Grove Stock Farm, in Nashua, Missouri. He served as president in 1905, the year the American Royal was incorporated, and as secretary in 1909. His grandson, John A. Dillingham, has pointed out the improbability that a Galloway man would be selected to head a cattle show dominated by Herefords and Shorthorns: It stands as further proof of the spirit of cooperation which prevailed at the Royal. From 1906 through 1912, Mr. Thompson was general manager of the American Royal. Later, he was manager of the horse show from approximately 1924 until the beginning of World War II.

Horse shows were popular events in Kansas City. In 1892, horse trainer Tom Bass organized a horse show as a fundraiser for the Kansas City Fire Department. Mr. Bass was the grandson of one of the richest men in Missouri, but because his mother was a slave, he was born into slavery in 1859. He trained champions including Miss Rex and Belle Beach, a filly who loved to waltz to "Turkey in the Straw." Mr. Bass designed the "Bass bit," an adjustable mouthpiece which is still in use, but he never patented the invention. For many years, he operated a stable at 39th and Main streets. At that location he started the Tom Bass Riding Clubs for young riders. The Kansas City Stock Yard Company later hired him to teach children to ride at the American Royal facilities.

The show became an annual event. It was held outdoors at Fairmount Park for several years

Allen M. Thompson (Photo courtesy of the Western Historical Manuscript Collection-Kansas City).

Tom Bass (Photo from *Tom Bass: Black Horseman* by Bill Downey).

and then moved to the Convention Hall. In 1896, Loula Long, then fifteen years old, won her first blue ribbon riding a reddish chestnut gelding named "Redbuck" in the class for the best lady rider. In 1902, the show at Convention Hall featured bronco-busting contests by cowboys from Idaho, Wyoming, Montana, Colorado and New Mexico. A program for the 1914 Kansas City Horse Show is in existence; it is possible (although not certain) that the show later merged with the American Royal, but the date is unknown.

There were 8,000 spectators on hand for the opening night of the first major horse show at the American Royal on October 17, 1905. Polo, then in vogue in the Midwest, was played the first evening. Draft horses had been shown at the American Royal in 1903, and Tom Bass had staged a special performance featuring Twilight, a five-gaited horse, and Limestone Belle, a high-schooled dressage horse in 1904, but this was the first real horse show at the American Royal. It is uncertain whether Loula Long was there: Records do show that she exhibited a horse named Shoo Shoo in 1906. Shoo Shoo failed to win on that occasion, but for more than fifty years thereafter, Loula Long Combs dominated the horse show at the American Royal.

Mrs. Combs was a child of privilege: the daughter of lumberman Robert A. Long, she grew up in a mansion named "Corinthian Hall" which now houses the Kansas City Museum. She wrote in her memoirs, "Daddy never refused to let me buy any horse I wanted, and I wanted many of them." But any competitor who expected to be up against a spoiled, easily distracted rich girl was in for a surprise: Mrs. Combs had talent and discipline to match her money and acquisitiveness. She bought the horses; together they won the ribbons.

Mrs. Combs' favorite horses were chestnut-colored with white stockings. She first rode saddle

13

Loula Long (later Combs) circa 1896 (reprinted in *The Independent* magazine).

horses, then switched to harness horses. In 1908 she acquired a horse she named "Sensation." From then on, all of her harness horses were given names which ended in "ion." Mrs. Combs also loved Boston bulldogs and often brought them into the arena with her. Audrey R. Peak recalled that: "Mrs. Combs was beloved by all, such a gracious person. When she drove into the ring, in her phaeton, with her two dogs in attendance, the crowd always gave her a standing ovation." Mrs. Combs was also known for her stalls at the American Royal: They had wood walls and were decorated with flowerpots.

Mrs. Combs' horse trainer for many years was a Scotsman named Dave Smith. John "Johnny" Haffey came to Longview in 1911. He broke horses, served as Mrs. Combs' footman when she drove phaetons at the Royal, and later showed horses, including Vibration and Little Abner. Mr. Haffey missed only two American Royals between 1912 and 1960. When he died on a Sunday night in 1966, his clothes were laid out for the next day: Mr. Haffey had planned to go to the Royal.

Dan D. Casement brought his first carload of Herefords from his Colorado ranch to the American Royal in 1907. Mr. Casement's background did not suggest that he would make cattle his career: He was a graduate of Princeton University, with a law degree from Columbia University. His father, General Jack Casement, owned Juniata Farm in Manhattan, Kansas. It was there that 16-year-old Dan saw his first registered Hereford. He later inherited the property from his father, and in 1915, moved from Colorado to Juniata Farm. Donald R. Ornduff wrote of him in *The First 49*:

> Although customarily seen in the battered garb of a working cattleman while engaged with the daily tasks at Juniata, he took delight in wearing the somewhat flamboyant and picturesque attire which became his virtual trademark when at the stock shows. With his spectacles attached to a wide black ribbon and his pipe almost ever-present, he was the picture of a legend when he donned his suit of black and white shepherd's check, his brilliant weskit of Stuart plaid, and his red tie.

Bud Snidow recalled of him: "Dan was the most complete cusser I have ever known, and I've known more than just a few." Mr. Casement won over 300 ribbons at stock shows in his career. He took home blue and purple ribbons from the 1952 American Royal. Mr. Casement died in 1953.

Many new events were introduced at the Royal in its early years. Some did not catch on, but others stayed and became part of the tradition. A flower show was held in 1907. Goats were part of the program for several years, including 1902 and 1907. Someone let a zebra loose in the mule show of 1907. Pandemonium ensued as the frantic mules dragged their handlers around the arena. Poultry was shown for the first time in 1908.

As the show grew, it became more difficult to find space to accommodate it. In 1908, a one-story stucco building was erected by the Kansas City Stock Yards Company at 20th and Genessee streets. This arena, where the judging of livestock took place in the daytime and horse shows were held at night, was called the American Royal.

The reaction of children to events at the Royal was already a source of amusement to their elders, as is shown in this anecdote printed in the *Kansas City Star* on October 13, 1908:

A small young lady was among the spectators in the "big top" yesterday. With intense interest she watched the great cattle led into the ring by their attendants. Each of the men who led the entries wore placards, sandwich-man fashion, showing the number of the animal entered, its age and weight. The little girl puzzled over the cards for some time. At last turning to her father she said, in a loud whisper: "Papa, those men must have got their cards all mixed. That little man there must be more than two years old and he can't possibly weigh 1,430 pounds."

Dan D. Casement and Donald R. Ornduff (taken at the 1946 Royal by the *Hereford Journal*; from *The First 49* by Donald R. Ornduff, used with permission of the American Hereford Association).

Estimated attendance swelled from 62,000 in 1908 to 86,000 in 1913. Rural dwellers came by train to the city. Out-of-towners as well as area residents began turning up for the show. The *Kansas City Times* recorded on October 7, 1913: "One of the noticeable features of the show is the attendance of strangers, whose dress and characteristics indicate that they are from the West. The broad brimmed hat is much in evidence, and everywhere yesterday the air of the West prevailed." Most of the 1914 show was canceled due to an outbreak of foot and mouth disease. There was a horse show that year, but no cattle. Thomas Wornall served as general manager in both 1913 and 1914.

In 1915, the American Royal moved its horse show to Convention Hall. This arrangement lasted two years. The carlot division was exhibited at the stock yards. Prize ribbons for the 1915 cattle show are lettered in gold "19" and "15" with a gold box between the two sets of numbers. Presumably, these are the unused ribbons from the show canceled the previous year: The numeral "14" is no doubt hidden within the box.

Robert Hazlett had begun exhibiting Herefords at the American Royal in 1909. His El Dorado, Kansas farm was named Hazford Place, Hazford being a combination of Hazlett and Hereford. Mr. Hazlett served as president of the American Royal in 1912, 1916 and 1922. In 1915, his entries won three blue ribbons. The following year, one of these winners, Bocaldo 6th won the senior and grand championship award. Mr. Hazlett continued to show Herefords until shortly before his death in 1936. The 1933 champion bull, Zato Rupert, and the 1933 champion heifer, Iza Rupert, were full brother and sister from the Hazford Place herd.

The 1916 show was marked by the appearance of Longhorn cattle. The *Kansas City Star* described them as "the very antithesis of the prize winners and the remnant of a wild, fierce, almost extinct race of cattle." Popular events included the "frontier contests," such as roping, riding, bull-dogging and "steertying." This was not for men only: Lucille Mulhall, a famous cowgirl of the period, was one of the competitors. On October 2, 1916, the *Kansas City Star* noted:

Not only is the effect of better prices perceptible in the personal property of stock raisers here for the American Royal, but in the stock show itself. Exhibitors have come distances this year which they never came before. The far-off state of Washington is represented for the first time and by a good-sized herd. There are a number of new exhibitors, men who a few years ago regarded a stock show as instructive and

Robert H. Hazlett with Zato Rupert and Iza Rupert at the 1933 American Royal (photo by Guy E. Smith, from *The First 49* by Donald R. Ornduff).

entertaining, but far too expensive for their limited capital.

The show was succeeding both as entertainment and as a business-oriented event. The American Royal Live Stock Show catalog for 1916 listed the requirements for entrants in the judging contest:

Any young man under 25 years old, and any student in an agricultural college who is an under-graduate, and who has not taken part in any live stock judging contest of interstate or international character prior to 1916, and who has taken at least twelve weeks' under-graduate work during the calendar year in the college that he represents, and who has, at no time, served in the capacity of animal husbandry teacher in any agricultural college, is eligible to enter as a contestant.

W. L. Nelson of Columbia, Missouri, was in charge of the livestock judging during the 1910s and 1920s. Livestock judging was then and still is an important part of the Royal.

Much happened to the detriment of the Royal in 1917. Mules and horses, as well as men, were shipped overseas to the first World War. On October 16, 25 acres of cattle pens and 12,000 animals were destroyed in a fire at the Kansas City Stockyards. Damages were estimated at one million dollars for the livestock and $175,000 for the yards and buildings. The Hartford Insurance Company paid over one million dollars in claims; Jay B. Dillingham has stated that owners of the livestock lost in the fire were fully compensated for the animals. That year, the Royal was held at Electric Park, an amusement park, and no carlot division was shown. On a happier note, it was on June 30, 1917 that Loula Long married Robert Pryor Combs. That autumn, the American Royal honored the bride with a special "Miss Loula Long Night" which featured driving and riding classes ("with practically all the hitches that go to make a complete show") and a horse show by Mrs. Combs and her stable.

In 1918, Canadian Herefords were exhibited at the Royal for the first time. Although the poultry show was discontinued that year, a man dressed as a "dancing rooster" entertained the crowd.

The year 1919 saw the show's return to Convention Hall. It was clear that the Royal needed its own building.

The 1920s

After World War I, General John Pershing's horse was stationed at Fort Leavenworth Cavalry Headquarters. The horse was exhibited at the American Royal by Col. Pete Carpenter.

In 1920, plans were drawn for a permanent home for the American Royal. Costs were estimated at $500,000. According to Jay B. Dillingham, Allen Thompson went to Chicago that year to meet with Thomas Wilson, who later owned the Wilson Packing Company. At that time, Mr. Wilson was in charge of the Morris family interests. The Morris family owned Morris and Co., which had a packing plant in Kansas City, and was the largest shareholder in the Kansas City Stock Yards Company, which owned the land where the American Royal building was to be erected. Morris and Co. owned a large herd of cattle in the Southwest. The company bought carloads of bulls — sight unseen — in the Midwest and shipped them by rail to join the herd. Mr. Thompson waited outside Mr. Wilson's door for an hour. As Mr. Dillingham has told it, "Finally, Mr. Wilson came out and said, 'Well, you can go home now, we've decided to build the new American Royal Building for you.'"

The Pereda Cup, which was awarded to the Hereford exhibitor who had bred and owned the three best bulls at the American Royal, was retired in 1920 after it was won in three consecutive years by O. Harris & Sons of Harris, Missouri with Repeater Jr., Repeater 129th and Repeater's Model. Repeater Jr. was the grand champion bull in 1918, 1919 and 1920. His record remains unbroken, and because the rules of the event have changed, no bull is likely to challenge it. The cup is four feet tall and is made of solid silver. As of 1999, it is in the possession of Andy Harris of Omaha, Nebraska, who is a grandson of O. Harris.

On November 7, 1921, the cornerstone was laid for the first American Royal building. The building was dedicated November 19, 1922. Governor Arthur Hyde of Missouri and Governor Henry Justin Allen of Kansas attended the ceremony, at which the Lindsborg Chorus of Lindsborg, Kansas sang the "Messiah." The Kansas City Star noted that: "The floor area of the new American Royal home 7-1/2 acres, larger by 1-1/4 acres than the International Amphiteater in Chicago, the building in which the International Livestock Exposition is held, one of the largest structures of its kind in the world." The estimated cost of the building was approximately $800,000.

A.M. "Andy Pat" Paterson came to the Royal in 1923 — and stayed 42 years. Mr. Paterson was born in Scotland in 1886. He had been head of the sheep department at Kansas State University prior to being hired by the Kansas City Stockyards Company to serve as assistant secretary of the

A. M. "Andy Pat" Paterson stayed with the Royal for 42 years in such positions as assistant secretary, livestock show manager, and general manager (Photo by permission of the American Hereford Association).

Left to right: Vern Elizabeth Purdy, Fairview Buster and Mabel Purdy in 1923. The girls' father, Grant W. Purdy was a well-known Shorthorn breeder from Harris, Missouri. Fairview Buster, a steer shown by Mabel Purdy, took first place in the first Boys and Girls 4-H Club Department. His twin, Fairview Emblem, shown by Vern Elizabeth Purdy, placed second (Photo courtesy of the American Royal Association, Vern Purdy Meek, and Western Historical Manuscript Collection-Kansas City).

Best 10 Herd, 1923 (Photo courtesy of the Western Historical Manuscript Collection-Kansas City, American Royal Collection).

Royal. Frank H. Servatius was secretary at that time, and served as general manager from 1922 through 1937. Mr. Paterson was appointed livestock show manager in 1938, and also served as general manager from 1941 to 1952 and as secretary from 1946 to 1952. His wife, Janet, helped put together the premium lists. Andy Pat was always chewing tobacco, and he liked to greet people by hugging them. This meant that they were splattered with tobacco juice, but his friends still laugh at the memory. Mr. Paterson retired to his farm in Iowa in 1965.

By 1923, the American Royal was such a fixture of Kansas City that a pundit for *The Independent* magazine could reflect on its earlier glories:

> Tuesday night marks the opening of the Horse Show. What memories of former smartness it brings to mind! I look back and visualize the Vanderbilt box, with the modest Reginald Vanderbilts, not so good looking nor so sartorially smart as our home folks . . . How our eyes would strain to see the first entrance of Loula Long

Horse parade, 1924 (Photo courtesy of the Western Historical Manuscript Collection-Kansas City, American Royal Collection).

Combs in her stunning turnouts and how feminine hearts fluttered when Ferdie Hornbeck brought out Gaity Girl and dear handsome Tom Velie would dash in on his polo pony . . . Even old Tom Bass and his High School horse gave us a thrill . . . Let's drop our blase attitude towards amusement, and display some of our old time enthusiasm for the cleanest and most delightful of entertainment — the Horse Show.

As so often happens in history, progress bred nostalgia.

At Harold W. Thieman's first trip to the American Royal in 1924, his Shorthorn steer placed second to last. His brother's steer was last. "We had a very poor judge. My brother thought so too!" he joked. In the 1920s, the Thieman family of Concordia, Missouri were among the cattle breeders who traveled the circuit from June or July until the beginning of December. They would take a carload (usually 15 to 17 head) of cattle by railroad to shows in Missouri, Iowa, Illinois, Kansas, Oklahoma and Nebraska. In order to get their Shorthorns to the American Royal, the Thiemans would rent a 50-foot railroad car and outfit it on the tracks at Higginsville, Missouri. "We would take out the third and the fifth board on each side. We could tie the cattle in, and it would give them air. We would build a deck right across the top of it on each end. We'd put our feed, bedding and equipment on one end, along with a tank that would hold a hundred gallons of water," Mr. Thieman recalled. After the American Royal, there was a show in Wichita, followed by the Chicago International. At the end of the season, the cattle would be delivered to the buyers.

Mr. Thieman especially enjoyed his participation at the Royal and in the 4-H club: "One of the highlights of being a 4-H member at the Royal was the tours the Chamber of Commerce afforded us. We were taken to the Loose-Wiles Biscuit Co., the *Kansas City Star*, Swope Park via streetcar and on buses to Longview Farms. My father, my brother and I had a room at the Coates House Hotel, and

Boys and Girls 4-H Club Calf Group, 1924 (Photo courtesy of the Western Historical Manuscript Collection-Kansas City, the American Royal Association, and Harold W. Thieman).

Fire damage, 1925 (Photo by Anderson, courtesy of Western Historical Manuscript Collection-Kansas City, American Royal Collection).

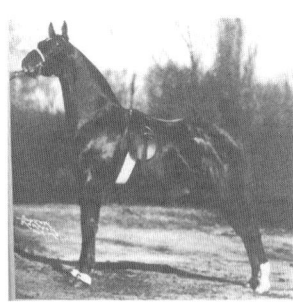

Chief of Longview (From *My Revelation* by Loula Long Combs).

a few years later, we stayed at the Commonwealth Hotel. We rode the streetcar on 12th Street to the Royal." Mr. Thieman later bred polled Shorthorns. He served on the Board of Directors of the American Royal and was chairman of the livestock committee. His brother, Homer Thieman was the editor of *The Daily Drover's Telegram.*

The roof of the American Royal building caught fire and caved in during an automobile show on February 13, 1925. Supporters of the Royal banded together, and the structure was fully restored in time for the November show.

Chief, a copper-colored chestnut bred at Loula Long Combs' stable at Longview Farms and trained by Lonnie Hayden, won the five-gaited stake at the 1925 American Royal. The following day, Chief was purchased for Lurlene Matson Roth (Mrs. W. P. Roth) by her mother. Chief was retired at the American Royal in 1932.

"Kansas Day" and "Missouri Day" both began in 1926, as did the American Royal Parade. From 1926 until 1946, George Catts served as parade chairman. Parades in the early years consisted of a few bands and horse-riding patrols from the stockyards. Floats decorated with real flowers were a highlight of the 1940s; in the 1950s, paper flowers were used. Parade chairmen have included Walter Atzenweiler, Richard Challinor, Leo Brady, Jr. and John B. Gage II.

For some, the highlight of the 1926 show was the fact that Will Rogers was there. More than seventy years later, Clint Tomson, whose uncle, John R. Tomson, was American Royal president in 1926, would write of Will Rogers: "[The] biggest honor I have ever been accorded was to be his 'Attendant' . . . I remember he used my younger brother, who was the same age as his youngest son, to try out a pony he purchased at the American Royal which was easily the proudest moment my brother, Jim, ever had." Other Kansas Citians were most enthralled about the appearance of real royalty at the Royal. On Armistice Day, 1926, President Calvin Coolidge, with Mrs. Coolidge looking on, dedicated the Liberty Memorial. Queen Marie of Rumania and her son, Prince Nicholas, and daughter, Princess Ileana,

The American Royal: 1899-1999

The 1926 American Royal Schedule

This schedule illustrates how the event had grown since the 1899 National Hereford Show. "Fat livestock" was the term used to designate market-ready animals. In 1926, the fee for general admission was $0.25; it doubled two years later.

Friday, November 12

9:00 am	Boys' and Girls' Club Judging Contest.

Saturday, November 13

8:00 am	College Student Judging Contest.
9:00 am	Opening of Livestock Show with full exhibit of Cattle, Hogs, Sheep, Horses and Mules.
9:00 am	Opening of Industrial Exposition and Manufacturers' and Jobbers' Industrial on second floor main building; also Farm Machinery and Implement Exhibit in First Floor Annex.
10:00 am	Cat Show, open all day and evening.
1:00 pm	Judging Boys' and Girls' 4-H Livestock. All classes.
3:00 pm	Special Children's Entertainment in Arena.
8:00 pm	Horse Show Classes; Spectacular Artillery Drill, Band Music, Gymkhana Races and Others. Special Features.

Sunday, November 14

9:00 am	All Livestock Exhibits open for inspection.
3:00 pm	Special Musical Festival.

Monday, November 15

8:00 am	Vocational Agriculture Judging Contest.
9:00 am	Judging Individual Fat Steers. All classes.
9:00 am	Judging Individual Fat Swine. All classes.
9:00 am	Judging Individual Fat Sheep. All classes.
10:00 am	Cat Show, open all day and evening.
11:00 am	Judging Hereford Breeding Cattle.
11:00 am	Judging Shorthorn Breeding Cattle.
11:00 am	Judging Angus Breeding Cattle.
1:00 pm	Judging Hereford Breeding Cattle.
1:00 pm	Judging Shorthorn Breeding Cattle.
1:00 pm	Judging Angus Breeding Cattle.
1:00 pm	Judging Individual Fat Swine. All classes.
1:00 pm	Judging Individual Fat Sheep. All classes.
2:00 pm	Judging Dorset Sheep.
7:30 pm	Million Dollar Parade of Prize-winning Livestock.
8:00 pm	Horse Show: Harness Horses, Roadsters, Saddle Horses, Hunters and Jumpers, Shetland Ponies, Hackneys, Six-Horse Hitch, Spectacular Artillery Drill and Special Band Music.

Tuesday, November 16

9:00 am	Judging Percheron Horses.
9:00 am	Judging Mules.
9:00 am	Judging Jersey Cattle.
9:00 am	Judging Ayrshire Cattle.
9:00 am	Judging Shropshire Sheep.
9:00 am	Judging Spotted Poland-China Swine.
9:00 am	Judging Carlot Feeder Cattle.
12:00 noon	Judging Hereford Breeding Cattle.
12:00 noon	Judging Shorthorn Breeding Cattle.
12:00 noon	Judging Polled Shorthorn Cattle.
1:00 pm	Auction sale of Purebred Angus Cattle.
1:00 pm	Judging Southdown Sheep.
1:00 pm	Judging Poland-China Swine.
1:00 pm	Judging Carlot Fat Cattle.
1:00 pm	Judging Carlot Swine.
3:00 pm	Horse Show Matinee, seats free.
5:00 pm	Horse and Mule Pulling Contest.

The 1920s

7:30 pm	Million Dollar Parade of Prize-winning Livestock.
8:00 pm	Horse Show: Roadsters, Harness Horses, Three and Five-Gaited Saddle Horses, Hunters and Jumpers, Shetland Ponies, Hackney Horses, Six-Horse Teams, Spectacular Artillery Drill and Special Band Music.

Wednesday, November 17

9:00 am	Judging Percheron Horses.
9:00 am	Judging Mules.
9:00 am	Judging Jersey Cattle.
9:00 am	Judging Holstein Cattle.
9:00 am	Judging "B" type Rambouillet Sheep.
9:00 am	Judging Hampshire Sheep.
9:00 am	Judging Berkshire Swine.
10:00 am	Dog Show, open all day and evening.
12:00 noon	Judging Hereford Cattle.
12:00 noon	Judging Angus Cattle.
12:00 noon	Judging Galloway Cattle.
1:00 pm	Auction sale of Purebred Shorthorn Cattle.
1:00 pm	Judging "C" type Rambouillet Sheep.
1:00 pm	Judging Oxford Sheep.
1:00 pm	Judging Chester-White Swine.
3:00 pm	Horse Show Matinee, seats free.
5:00 pm	Horse and Mule Pulling Contest.
7:30 pm	Judging Best Ten Head Herefords, Shorthorns, Angus and Galloways for Stockyards Trophies.
8:00 pm	Horse Show: Roadsters, Harness Horses, Three and Five-Gaited Saddle Horses, Hunters and Jumpers, Shetland Ponies, Hackney Horses, Six-Horse Teams, Spectacular Artillery Drill and Special Band Music.
8:00 pm	Dog Show.

Thursday, November 18

9:00 am	Judging Belgian Horses.
9:00 am	Judging Clydesdale Horses.
9:00 am	Judging Milking Shorthorn Cattle.
9:00 am	Judging Guernsey Cattle.
9:00 am	Judging Hampshire Swine.
9:00 am	Judging Cotswold Sheep.
9:30 am	Auction sale of Carlot Fat Cattle. Sale of Carlot Feeder Cattle to follow immediately.
10:00 am	Dog Show, open all day and evening.
11:00 am	Auction sale of Fat Sheep.
12:00 noon	Judging Shorthorn Cattle.
12:00 noon	Judging Polled Shorthorn Cattle.
12:00 noon	Judging Guernsey Cattle.
12:00 noon	Judging Milking Shorthorn Cattle.
1:00 pm	Auction sale of Purebred Hereford Cattle.
1:00 pm	Judging Duroc Swine.
1:00 pm	Judging Lincoln Swine.
1:00 pm	Auction sale of Fat Carlot Swine.
2:00 pm	Auction sale of Individual Fat Swine.
3:00 pm	Horse Show Matinee, seats free.
5:00 pm	Horse and Mule Pulling Contest.
7:30 pm	Million Dollar Parade of Prize-winning Livestock.
8:00 pm	Horse Show: Roadsters, Harness Horses, Three and Five-Gaited Saddle Horses, Hunters and Jumpers, Shetland Ponies, Hackney Horses, Six-Horse Teams, Spectacular Artillery Drill and Special Band Music.
8:00 pm	Dog Show.

The American Royal: 1899-1999

Friday, November 19

9:00 am	Judging Belgian Horses.
9:00 am	Judging Shire Horses.
9:00 am	Judging Shorthorn Cattle.
9:00 am	Judging Hereford Cattle.
9:00 am	Auction sale Boys' and Girls' 4-H Club Calves.
10:00 am	Dog Show, open all day and evening.
1:00 pm	Judging Hereford Cattle.
1:00 pm	Judging Shorthorn Cattle.
1:00 pm	Auction sale of Fat Steers.
3:00 pm	Horse Show Matinee, seats free.
5:00 pm	Horse and Mule Pulling Contest.
7:30 pm	Million Dollar Parade of Prize-winning Livestock.
8:00 pm	Horse Show: Roadsters, Harness Horses, Three and Five-Gaited Saddle Horses, Hunters and Jumpers, Shetland Ponies, Hackney Horses, Six-Horse Teams, Spectacular Artillery Drill and Special Band Music.
8:00 pm	Dog Show.

Saturday, November 20

10:00 am	Dog Show, open all day and evening.
10:30 am	Grand Parade of Prize-winning Beef Cattle.
11:00 am	Grand Parade of Prize-winning Dairy Cattle.
11:30 am	Grand Parade of Prize-winning Draft Horses and Mules.
2:30 pm	Horse Show, 3,000 free seats.
8:00 pm	Horse Show: Roadsters, Harness Horses, Three and Five-Gaited Saddle Horses, Hunters and Jumpers, Shetland Ponies, Hackney Horses, Six-Horse Teams, Spectacular Artillery Drill and Special Band Music.

arrived in town that day and attended a concert at the American Royal Pavilion. *The Independent* recorded: "Queen Marie's gown was of pearl encrusted openwork embroidery over black and she wore a headdress of pearls and diamonds." In those days, people dressed in evening clothes for the horse shows. During the late 1920s and early 1930s, the chairs at the American Royal had royal blue linen seat covers.

Because their father was the horse show manager, Louise and Frances Thompson often brought their beaus to the show. It was in the old American Royal Arena that Frances had her first date with a young lawyer named Jay B. Dillingham. The couple courted for several years and were married on September 28, 1935. Mr. Dillingham went to work for the Kansas City Stock Yards Company in 1937. At that time, George R. Collett was president of the company and W. H. "Billy" Weeks was vice-president; both were very much involved in the American Royal. Mr. Weeks was general manager of the Royal from 1916 through 1921 and again in 1938 and 1939. Mr. Dillingham became president of the company in 1948 and later served as president of the American Royal.

It is a oft-repeated fact that the first national convention of the Future Farmers of America was held in Kansas City in 1928. Less well-known is the fact that the FFA (then known as Smith-Hughes agricultural students, because the organization was created by the Smith-Hughes Act in 1917) held livestock judging contests at the American Royal in 1926 and 1927. This was very definitely a home-grown activity: Among those who helped create this contest were officials from the American Royal, the Kansas City Stock Yards Company, the Kansas City Livestock Exchange, the *Kansas City Star* and the Chamber of Commerce of Kansas City, Missouri. Will Rogers wrote in praise of the students (and less kindly of Queen Marie):

I followed Marie, into Kansas City . . . Society has been burying [sic] dresses and scrubbing their necks and ears for weeks getting ready to see the Queen. The Mayor of Kansas City in presenting her said, "It's the greatest day in the history of Kansas City.

Truck Show, mid-1920s. Note biplane in background. (Photo by Anderson, courtesy of Western Historical Manuscript Collection-Kansas City, American Royal Collection).

Now can you imagine such a statement. Just take it apart and see . . . Why last week when I was there, there were 17 hundred young boys and girls brought there by that great Paper, the Kansas City Star, from over 30 states. They were taking vocational training and had led their various districts back home in the studying of farming, and stock raising, and had been brought to see the American Royal Live Stock Show. To see the Kings and Queens of Cattle, Sheep, Hogs, Horses. Real Kings and Queens that produced something . . . These not only have to have the breeding, but they got to face the Judges and be marked on their merit.

There were no "bluejackets" in Will Rogers' day. The dress code that mandated blue corduroy jackets with gold trim was enacted in 1952. The uniforms gave the organization an identity: The swarm of "bluejackets" would be instantly recognizable in the crowds at the American Royal, as well as in hotels and stores and on the streets of downtown Kansas City.

In 1927, Charles A. Lindbergh made the first solo, nonstop transatlantic flight; that year Audrey R. Peak attended her first American Royal horse show as the bride of J. Ralph Peak. His father, George J. Peak, and grandfather, J. R. Peak, had shown roadsters and Shorthorn cattle in the early days of the Royal. By the late 1920s, the stable of George J. Peak and Sons, (J. Ralph and his brother Mark C. Peak), was exhibiting roadsters, heavy harness horses and Hackney ponies. "We would have around 20 head of horses, which we shipped by rail," Mrs. Peak later wrote. "[We would travel with] two railroad cars with horses, one car with tack and buggies, and six or seven grooms." J.

Johnie Miller at the Concert for Champions (Photo © by Howard Schatzberg, reprinted with permission).

Ralph Peak served as horse show manager from 1972, when he succeeded Bob Leu, until his retirement following the 1978 show.

James C. Swift served as American Royal president from 1927-1937. He was the president of the Swift & Henry Commision Company. "Jim Swift was one of the most eloquent speakers I ever heard," Jay Dillingham recalled.

In 1929, W. A. Cochel, managing editor of the *Kansas City Star*, announced that the newspaper would offer an annual prize of $1,000 to the Star Farmer of America. This award would go to an FFA member who demonstrated outstanding achievements in farming and leadership. The *Kansas City Star* ran the contest for 20 years before turning it over to the FFA at the FFA's request.

Three years after the first parade, John E. "Johnie" Miller began his unrivalled streak:

I rode in the American Royal Parade in a car as President of Club Presidents Round Table in 1929.
    Then in 1930 I started riding in the American Royal Parade with the Mounted Guard of the Shrine each year to 1941 — when I changed over to the Saddle & Sirloin Patrol where I have ridden ever since. 1989 was my 60th American Royal Parade riding a horse, except that [in 1989] as Grand Marshal I rode in an English wagon drawn by draft horses.

A picture of Johnie Miller leading the parade hangs on the wall in the American Royal Museum. He liked to visit it, and he would say, "That looks just like me — except he's too gray."
    Ray Davis attended his first American Royal in 1929, but he hoped no one noticed: The teenager had snuck in by climbing up the fire escape. Mr. Davis, who grew up on a farm in Romance, Missouri, spent the next 57 years in the cattle business. He served as superintendent of the steer sale at the Royal for many years, and was a member of the board of directors. In 1983, he headed a group of investors that bought the Kansas City Stock Yards. He continued to buy cattle until a week before his death in 1991.

Ray Davis with FFA member and prize-winning sheep, 1978 (Photo courtesy of the Western Historical Manuscript Collection-Kansas City, American Royal Collection).

    1929 was also the year that a young couple from Oklahoma named Gladys and Joe Mackey first came to the Royal. The Mackeys had just moved to Kansas City, where Mr. Mackey trained horses at the R. L. Nafziger's Somerset Place farm. Mrs. Mackey later recalled that the arena was sold out the first night they were there, and they stood and watched the show. For years at the Royal, Mr. Mackey was a familiar presence with his Western hat, his cigar and his cane. He taught children's riding classes, and his students still remember the sound of his cane rapping against the railings. His niece, Bev Chester, assisted him in teaching. Kenny Burgdorfer, a longtime exhibitor at the Royal, was among his students. When Mr. Mackey died in 1976, Mrs. George Bunting said of him, "He was the most influential man in this part of the country in hunters and jumpers and in giving encouragement to children."
    The Mackeys' son, Joe, Jr., made his first appearance at the Royal at the age of three. He won a children's pony event in 1935, when he was five years old. He had collected 87 trophies from the Royal and other horse shows by the time he was 13. He was also accomplished polo player prior to his death from a cerebral hemorrhage at the age of 21. His sister, Patsy Mackey (later Patsy Hahn), rode in many American Royals, and her daughter, Sherri Hahn Kahn, has continued the Mackey family

Joe Mackey (Photo courtesy of Gladys Mackey).

James Lee with his purchase, 1929
(Photo courtesy of Col. and Mrs. John
Riffle).

Livestock judging, circa 1920s (Photo courtesy of
the Western Historical Manuscript collection-Kansas
City, American Royal Collection).

Tarkio Molasses and Feed exhibit at the American Royal, circa 1929 (Photo by Anderson, courtesy of the
Western Historical Manuscript Collection-Kansas City, American Royal Collection).

tradition. The Joe Mackey Memorial Scholarship Fund, given in memory of the late horse trainer, is awarded each year to the high point junior rider in junior hunter and equitation divisions.

During the late 1920s and early 1930s, the American Royal Riding Academy was held every Saturday morning in the old American Royal Arena. The school was run by Margaret McLaughlin, and the horses were provided by business men who had offices in the Livestock Exchange Building. Students participated in their own horse show as well as in the American Royal Horse Show.

There was turmoil in the stock market, but the 1929 American Royal was a festive time for Kansas City's young riders. The local boys and girls jumping class was won by W. W. Guernsey on Rob Roy. Carol Hagerman (later Carol Durand) and Babe received the red ribbon, and James Kemper and his gray mare, Kitty, placed third. Carol Hagerman, who first competed in the Royal in 1925, won the local junior saddle class for girls in 1929. In describing her victory, the *Kansas City Star* also noted that "[her] riding distinguished more than one class last year." In addition, Miss Hagerman placed first in horsemanship. For that category, she rode a brown gelding named My Jimmy. Burleigh Wolferman was second; her horse was Bonnie Best, a chestnut mare. Ruth Dierks finished third aboard Beauty, a gray mare.

The American Royal grand champion Hereford steer of 1929, shown by Andy Harris of Harris, Missouri, sold for $3.00 per pound to James Lee, owner of the Electric Grill & Waffle Shop. Mr. Lee's son, Don, is the owner of the Savoy Grill, which features a "4-H Steak" on the menu.

The National FFA Band in an American Royal parade (Photo courtesy of the American Royal Association).

Saddle & Sirloin Club riders in an American Royal parade (Photo courtesy of the American Royal Association).

The Royalettes in an American Royal parade (Photo by Anderson, courtesy of Western Historical Manuscript Collection-Kansas City, American Royal Collection).

Parade Scenes Over the Years

Left to right: BOTARs Susan Bagby (now Biggar), Jane Bredberg (now Nelson), and Julie Gempel (now Lindstrom) at the 1954 American Royal parade (Photo courtesy of the BOTAR Organization).

The 1959 American Royal Parade featuring Mary Diane Arnett, American Royal Queen (Photo courtesy of the American Royal Association).

The 1930s

W. C. "Bill" Theis first visited the American Royal with his Bryant School class on a field trip. Mr. Theis later headed the Simonds-Shields-Theis Grain Company and served as president of the American Royal from 1973-1974. Of his childhood visits, he later wrote, "I enjoyed all of the exhibits including farm animals as well as special exhibits of chickens, turkeys, rabbits and guinea pigs. The mules at the American Royal were a big attraction."

Mules were for many years a part of the American Royal. In the 1930s, a mule named Dixie dominated the competition, winning at least three grand championships and appearing on the front page of the *Kansas City Star*. "You have to crack whips when you are showing mules," a handler told the *Kansas City Star*. "Now you take those fancy show horses and when the crowd cheers and applauds, they do their stuff, step high and show off.

Red, champion mule, 1927 (Photo courtesy of Mildred (Mrs. James) Randall and Col. and Mrs. John Riffle).

"A mule is different. He kind of has an idea he was dragged in by some man and will be dragged out again. The only thing that makes an impression on a mule in the show ring is a loud whip crack. Then he shows some interest in just how close that whip is coming to Mister Mule."

Ed Frazer of Drexel, Missouri frequently exhibited his mules. On at least one occasion, he brought a 10-mule hitch to the Royal. More than fifty years later, John Riffle recalled with pleasure being one of the boys chosen to ride in the wagon for ballast. One of Mr. Frazer's animals found lasting fame in Hollywood: Francis the talking mule. Another mule from Mr. Frazer's stables, Samson (also known as Champ Clark) appeared with Bob Burns in the film "I Am from Missouri." Mr. Burns was so taken with the mule that he purchased Samson at the conclusion of filming. Mr. Frazer's American Royal grand champions included Maud and Jane. The last mule competition at the Royal was held in 1955.

The draft horse competition was another prominent feature of the American Royal until the end of the 1930s. Draft horse judges included Dr. Don Kays of Ohio State University, L. P. McCann of Ohio, Wayne Dinsmore of Illinois, Dean Eddie Trowbridge of the University of Missouri, E. B. White of Virginia, Jerry Moxley of Kansas and Frank J. Rathje.

> Here is a snapshot of the stockyards from *The WPA Guide to 1930s Missouri*:
> [the stockyards] which extend from Twenty-Third Street on the south to Twelfth Street on the north, and from Genessee Street to the Kaw River, are 238 acres of pens connected with the large packing house by chutes and runways and a special double-decked bridge. Of the yardage, 87 acres are under cover. Daily capacity is 70,000 cattle, 50,000 hogs, 50,000 sheep, 5,000 horses and mules. Horse and mule auctions are held each Monday in a barn at the southeast corner of Nineteenth and Wyoming Streets. At Twenty-Third Street is the American Royal Pavilion, erected to house the American Royal Horse and Livestock Show . . . The present building dates from 1925, when its predecessor was destroyed by fire.

Donald Ornduff later recalled that in the 1930s it was relatively inexpensive for exhibitors to come to the American Royal, partly because labor costs were low. Many Scots were employed as stockmen; some had traveled to America as shipboard attendants for the animals. "It didn't cost much to bring the cattle in [to the show]. They would have some local boys and some full-time employees of the farm. To add a herdsman from Scotland didn't cost much. Some of the owners did a lot of the work themselves. In those times, most of the people — even in town — were only a generation or two away from the country." The herdsmen sometimes had 4-H participants help show the cattle. "I was lucky and the herdsman from Thomas Wilson's Edellyn Farms let me help show their Shorthorns. Each time, on leaving the ring, the herdsman would hand me a dollar bill. Back in the 1930s, that was worth something," recalled John Riffle, who grew up on a farm in Pleasant Hill, Missouri.

The 1931 American Royal Livestock Premium List contained the following caution for exhibitors:

> All hogs or barrows in the individual classes should preferably be shipped by express to yourself, care [of] American Royal Live Stock Show. If shipped by freight, care should be taken that notation, "Not to be sent to Stock Yards Chutes," is placed on billing and a caretaker should accompany same.

Sweetheart On Parade, which was trained by Lonnie Hayden and owned by Lurlene Matson (Mrs. W. P.) Roth's Why Worry Farm was the 1931 five-gaited champion, besting American's Dream, which was shown by Don Reavis for Carnation Farm. The champion was a favorite of the crowds: "Everybody would come to the Royal to see Sweetheart On Parade," John Riffle recalled. But this highlight of the American Royal were overshadowed by a train derailment. "The Million Dollar Train" took its name from the fact that it carried showhorses from the Royal to Chicago. Horses belonging to Anacacho Ranch, Bridgeford and Bradford, H. G. Eshelman, Mary Gwyn Fiers, R. C. "Doc" Flanery, Harry Gorham, Mary Liewellyn,

An American Royal Horse Show in the 1930s
(Photo by Anderson, courtesy of the Western Historical Manuscript Collection-Kansas City, American Royal Collection).

The 1930s

Hilda McCormick, E. Phillips Schandein and National Woolen Mills were aboard that afternoon. It had been raining for several days when the train left Kansas City. Tracks near Lexington, Missouri were washed out, and it was there that the wreck occurred just before 5 pm on Sunday, November 22. There were seven men killed: four grooms and three hoboes. Another 14 men were injured. Twenty horses died, including Mrs. A. C. Thompson's heavy harness horse, Scottish Rite. Pony trainer Lane Bridgeford later recalled leading two horses to the Lafayette County Home, where the survivors were taken. *The American Saddlebred* magazine quoted him as having said: "I didn't know I'd get to the poor farm with a pair of horses, but I always knew I'd get there."

The Golden Anniversary Hereford Show, which marked the 50th anniversary of the American Hereford Association, was held at the American Royal in 1932. Roland Jacob "R. J." Kinzer, who had served as secretary of the association since 1911, organized the show, which included 4,670 Herefords. Fred DeBerard of Kremmling, Colorado took home the blue ribbons for best yearling carload and for best carload of Hereford feeder calves. Not a bad showing, considering that it was Mr. DeBerard's first trip to the Royal. DeBerard calves would win many ribbons in the years to come.

In 1933, Lou Williams of the Williams Meat Company and Frank H. Servatius, general manager of the American Royal, created the Kansas City Strip Steak, which is also known as the Kansas City Boneless Strip Steak or the Kansas City Steak. The steak was thus named to honor the youngsters who brought their steers to the American Royal. James W. Leathers and Jay B. Dillingham later wrote this description of this cut of beef:

> a steak approximately one inch thick, cut from an oblong muscle of the loin only 8 to 9 inches wide. It is a boneless cut of meat with a thin layer of fat on one side. A 1,000-pound choice grade steer producing a 600-pound carcass of meat will yield only 24 one-pound K. C. Steaks. The meat comes from behind the 13th rib, where it is surrounded by some "mighty fine company" — the sirloin, T-bone, porterhouse and club steaks.

In 1998, controversy brewed over the two strip steaks when the Wall Street Journal ran a front page article in which Kansas Citians and New Yorkers disputed each other's claims to the boneless steak. However, the facts are these: the New York Strip Steak has a feather bone, and the Kansas City Strip Steak is boneless and far more flavorful (this is the official opinion of the American Royal).

That same year, Mr. Williams brought his nephew, Eddie, to his first American Royal. The younger Williams would later be known as "Mr. American Royal." Ever a promoter, Eddie Williams once made the national newspapers for sending Kansas City Strip Steaks on a TWA flight to the Kansas City Athletics baseball team when they were playing in Boston and were in a slump.

An activity that John Riffle greatly enjoyed as a youngster in the 1930s was the "Catch a Calf Contest": "Steer calves were turned loose in the arena and 4-H and FFA boys and girls tried to catch them. Those who caught calves were given the steer to raise, with the stipulation that they feed it and bring it back to the American Royal the next year."

President Franklin D. Roosevelt received a present from the 1934 FFA convention: a dressed lamb which had belonged to an FFA member prior to winning a prize at the Royal.

There was great excitement in 1935 when Sweetheart on Parade, owned by Mrs. W.P. Roth of Why Worry Farm and shown by Lonnie Hayden, competed against Carnation Hour, owned by Carnation Stables and ridden by Don Reavis. As a Kansas City Star reporter noted "Between Mrs. Roth's stables and Carnation Farms, Pomona, Calif, a keen rivalry had been built up in horse shows on the Pacific coast and elsewhere. That rivalry was brought to the American Royal, where Carnation was showing Carnation Hour, a big chestnut gelding, in a bid for the 5-gaited championship stake." Both

Best 10 Head, 1936 (Photo by Creswell's, Courtesy of Joan Edwards).

were outstanding horses, but that night Sweetheart on Parade bested Carnation Hour. Sweetheart on Parade had also won in 1931 and 1932, and therefore received not only the championship trophy, but also the *Kansas City Star* Challenge Trophy.

One horse stood out from the others in Kansas City in 1938 — and he wasn't even in the Royal. Lawrin, a brown colt belonging to Herbert Woolf, won the Kentucky Derby. Lawrin was buried on Mr. Woolf's farm, Woolford, in Prairie Village, Kansas. The farm has since been developed into a shopping and residential district. Hints of its past are presented in the names of the subdivisions: Corinth Paddocks and Corinth Downs. Lawrin's grave marker still stands in Corinth Square. Mr. Woolf was president of Woolf Brothers department store and was a strong supporter of the American Royal. He also bred Herefords. His nephew, Alfred Lighton, himself an accomplished horseman, later recalled going to the American Royal as a child, dressed in a tuxedo.

Betting is a common practice at the Kentucky Derby, but it is not permitted at the American Royal. Sometimes, however, that fact is ignored. The 1938 five-gaited championship, for example, apparently involved higher stakes than just prize money. Midnight Star (also known as Midnite Star) had won at Louisville, St. Louis and, most recently, at Ak-Sar-Ben in Omaha. He was considered the favorite to win, but his owner, Joe Freeman of Leisure Hour Stable, discovered that there was a tremendous amount of money bet against him. The Freeman family believed that William R. Skidmore was responsible. Mr. Skidmore, a Chicago gambler and an owner of Pine Tree Farm, had purchased a mare named Lady Jane from the dispersal sale of George Godfrey Moore's Georgian Court Stables earlier in the year. According to Amy Freeman Lee in her 1940 book *Hobby Horses*, the mare's owner first tried to corrupt a judge by offering to purchase one of the judge's horses at an inflated price. The offer was refused, but this was only the beginning. Mrs. Lee wrote:

> Grooms and others, whose price for their services was one dollar and a seat to the show, were hired and placed throughout the arena to applaud for the mare and to hiss the champion... The acts became more effective and at the same time more vicious. People were placed around the rail to "accidentally" drop hats or to make violent movements with their arms as the champion passed by in an effort to frighten him into making a mistake. The least subtle measure became apparent when dark horses were entered at the last minute in the championship stake. Certainly no one will spend several hundred dollars for an entrance fee in a class to show a horse which is totally incapable of doing enough to keep from looking ludicrous and incongruous. This meant only one thing — they would try to pocket the champion or knock him into the wall.

Mrs. Lee's fears were confirmed when she saw a rider deliberately hit Midnight Star's back. The blow did not effect his performance, and Midnight Star took the blue ribbon. Mrs. Lee took care not to mention Lady Jane or William Skidmore by name in her 1940 book; their identities were revealed in a 1981 *Saddle & Bridle* article written by Lynn Weatherman.

Lady Jane was born at Longview Farm; horse trainer Sug Utz has stated his belief that she was "probably the finest horse Loula Long Combs ever owned." Within a few years of the 1938 championship, Mr. Utz saw Lady Jane at a saddlehorse sale in Chicago. According to Mr. Utz, "Al Capone was supposed to have owned her when she was sold." Clifford Moore (no relation to George Godfrey Moore) from North Carolina bought Lady Jane for a brood mare. Mr. Moore paid $5000 for her, $9000 less than she had cost Mr. Skidmore.

In 1939, the first American Royal Queen was chosen. Margaret Jane Swift (later Margaret Fair) was a schoolteacher from Claremore, Oklahoma — the town that Will Rogers had made his home. Mrs. Fair later recalled that her most vivid memories were of "the man who seemed to be around all the time, talking all the time, making speeches at luncheons and everything else — H. Roe Bartle. He was certainly my good friend." Mr. Bartle was then an executive with the Boy Scouts. He served as mayor of Kansas City from 1955 to 1963.

The judges for the queen contests were Powell Groner, Ritchie Cooper and Elsa Maxwell. In honor of Miss Maxwell, a dressmaking contest was held. Contestants draping fabric on models included R. Crosby Kemper, Sr., J. C. Nichols, Roy Roberts and the winner, Harry Darby, who received a dozen neckties from Miss Maxwell. Originally, the American Royal Queen candidates were young women selected to represent cities in the Midwest. Civic organizations including Rotary Clubs, Elks, Chambers of Commerce and newspapers held contests. The winners came to the Royal. The primary qualifications for queen were "charm and personality," style (including "alertness to fashion trends") and, to a lesser degree, "culture" (education and talent). In 1961, the rules were changed: Midwestern colleges and universities selected and sent their candidates, and the girls represented their schools rather than their hometowns. College students gave way to high school girls in 1970, when the selection process was revised again. This time, the candidates were FFA Sweethearts chosen in statewide contests all over America.

Several American Royal Queens held other beauty contest titles. De Lois Faulkner, 1953 American Royal Queen, and Malinda Diggs Berry, 1957 American Royal Queen, served as Miss Maid of Cotton in 1954 and 1958, respectively. Sarah Kay Burns, 1960 American Royal Queen, was Miss

Missouri in 1961. Her successor, Carolyn Jane Parkinson, was Miss Kansas as well as American Royal Queen in 1961.

American Royal supporters and celebrities served as judges for the contest. Noteworthy judges included presidential candidate Thomas E. Dewey in 1948, baritone John Raitt in 1959, comedian Andy Griffith in 1962, actor Randolph Scott and modeling agency executive Eileen Ford in 1963. Actress Dorothy Lamour shared master of ceremonies duties with local television personality Ken Heady in 1957. Dinah Shore appeared at the Coronation Ball in 1969, when a group of Kansas City-area high school girls known as the Royalettes performed dance routines. In earlier years, the Royalettes served as pages and carried the queen's train. The Secret Service was on hand when one of President Ford's sons attended.

While no one can doubt that American Royal Queens were capable of far more than just looking good (as exemplified by Mary Jo Smith, 1956 American Royal Queen, who later achieved prominence as Dr. Mary Jo Evans, cancer researcher), the idea of a beauty pageant fell out of favor. The last American Royal Queen was crowned in 1988. The following year, the Student Ambassador program began.

Jay B. Dillingham, American Royal president, crowns American royal Queen Carolyn Jane Parkinson, 1961 (Photo courtesy of the American Royal Association).

American Royal Queen candidates, 1939 (Photo by Anderson, Courtesy of the Western Historical Manuscript Collection-Kansas City, American Royal Collection).

The 1940s

The 1940s were important years for the American Royal. This may seem surprising, given that the outbreak of World War II caused the Royal to abandon its usual schedule. Despite this fact, some of the events that would lay the foundation for the modern American Royal occurred at this time: In 1940, the Saddle & Sirloin Club was founded, Harry Darby became president of the American Royal in 1941 and the first Belles Of The American Royal made their bows in 1949.

Among the initial objectives of the Saddle & Sirloin Club set forth in its constitution were "to help improve and perpetuate the American Royal Livestock and Horse Show" and "to promote and improve livestock and agriculture in the territory tributary to Kansas City." The Saddle & Sirloin Mounted Patrol makes trips to other cities, including Santa Fe, Cheyenne, Fort Worth and St. Louis, to promote awareness of the Royal, and rides in the American Royal parade and other Royal events. The club also holds an annual horse show to benefit the American Royal. Many Saddle & Sirloin Club members chair Royal committees and serve as directors. American Royal presidents include 17 club members; eight men have headed both organizations.

Harry Darby, former U.S. Senator from Kansas, was the American Royal president from 1941 to 1952. He headed Darby Industries, a steel fabricating and manufacturing concern which built landing crafts for the Navy during World War II. In his spare time, Senator Darby raised Herefords, though he did not show them. For these reasons, he was sometimes referred to as a "Kansas industrialist and stockman," which amused him greatly. He served as Special Ambassador to Venezuela under President Eisenhower. Whatever else was on his agenda, the American Royal was a top priority. Senator Darby's pet project at the American Royal was "Kansas Day,"

Sen. Harry Darby and friends (Photo courtesy of the American Royal Association).

Saddle & Sirloin Club members on a trail ride in Santa Fe, New Mexico to promote awareness of the American Royal (Photo courtesy of the Saddle & Sirloin Club).

which was held on Wednesday in the years when the show lasted a week. "United States senators attended, representatives attended, the local politicians and business people came. It was a wonderful time," recalled Cindy Stanley, who was Senator Darby's secretary from 1973 until his death in 1987. For many years, a "Kansas Day" dinner was held at the Town House in Kansas City, Kansas. After dinner, guests would be bused to the American Royal. In later years, the dinner was held at the Reardon Civic Center.

Senator Darby was born in 1895, four years before the beginning of the Royal. He attended every American Royal from 1905 through 1986. His daughters, Joan Edwards, Edith Marie Evans, Marjorie Alford and Harriet Gibson and their families continue the Darby tradition of dedication to the American Royal. Two of Senator Darby's sons-in-law, Roy A. Edwards, Jr. and Ray R. Evans, served as American Royal presidents.

Sen. Harry Darby in 1941 (Photo by permission of KMBZ, courtesy of the Western Historical Manuscript Collection-Kansas City, American Royal Collection).

Roy A. Edwards, Jr. (Photo by Wilborn & Associates, courtesy of the American Royal Association).

Ray R. Evans (Photo by Wilborn & Associates, Courtesy of the American Royal Association).

Alberta Lee Cox, 1991 (Photo courtesy of Alberta Lee Cox).

The Roy A. Edwards, Jr. Memorial Grand Prix, which draws some of the finest horse trainers in the country, is held annually in memory of Mr. Edwards, who died in 1987. The Grand Prix was initiated by Dwight Sutherland and the BOTAR Organization. Mr. Edwards served as president of the Royal in 1977-1978; during his tenure, the Royal was awarded the United Professional Horsemen's Association "Best Show of the Year" citation. Mr. Evans, an All-America football player in 1947 and an All-America basketball player in 1942 and 1943, was president of the Royal in 1987-1988. The American Royal Center is dedicated to the memory of Senator Harry Darby and R. Crosby Kemper, Sr.

One of the big events of the 1941 show was the retirement of Loula Long Combs' 11-year-old white-legged mare, Captivation, who had been champion at Madison Square Garden in New York, at the World's Fair Horse Show held in 1939 on Treasure Island near San Francisco and at the Royal Winter Fair in Toronto. Mrs. Combs wrote in her book, *My Revelation*:

We decided that the harness stake would be Captivation's last show in competition and she would be retired on Saturday night... When the gates swung open for Captivation to make her last appear ance in the show ring, every seat in the large building was filled, and people were standing crowded aroundthe promenade. You could hear many saying, "Here she comes. "Cappie and I were both nervous, and she didn't settle until we had gone around the ring a couple of times . . . The applause she received would be music to the ears of any performer.

After a proclamation was read by Charlie Green, the horse show an nouncer, a blanket of flowers was put over Captivation. Mrs Combs noted, "Many people came to the stable to see the mare once more and to take a flower from her floral blanket."

Just beginning her career at the American Royal was Alberta Lee Cox, the young daughter of Leroy Cox who owned Blue Ridge Stables. Ms. Cox celebrated 50 years of showing horses at the Royal with a party in 1991. Horses she has shown include My Pretty Girl in the five-gaited amateur stake and, more recently, Commander's Mary Lou in the fine harness division. She recalls with pleasure seeing the champion horses, Emerald Future, shown by Lloyd Teater, and Meadow Princess, ridden by Lloyd's brother Earl. Another horse she greatly enjoyed watching was the harness show pony Holiday Spirit, which belonged to the Costellos of St. Louis, Missouri. "Everyone wanted to own Holiday Spirit — not just me," she said, laughing at the memory.

Humorist Calvin Trillin has written that his grade school classes went to the stockyards so often that he finally asked, "Please, teacher, can we have some arithmetic?" But he also has admitted that his older sister, Sukey Trillin Fox, said they never went to the stockyards. The explanation for this discrepancy may be the fact that in 1942 and 1943, the Royal held only a three-day fat livestock show (this would now be called a market livestock show). No American

The American Royal: 1899-1999

Royal was held in 1944, and the 1945 show was truncated. Mr. Trillin's teachers may have been excited because the war was over and the American Royal was back in full swing.

During the war, the building was leased for wartime industrial production to a company making military gliders. "Of course they took out everything in there and manufactured a lot of gliders. They'd take them down the street in pieces, take them to Fairfax and get them out on boats," Jay Dillingham recalled. "But after the war, we got into quite an argument with Uncle Sam about restoration costs. We gave the government an estimate of what we thought it would take to put it back like it ought to be. A young lieutenant from New Jersey who was stationed at Wright Field just couldn't believe what it would cost. He turned down the proposition and invited us to come and see him. The engineer and I went. We went all through it, piece by piece. When we got through, he said, 'Well, I understand that now.' I said, 'Fine. Now total it up.' It was quite a bit higher than our first offer. But he was happy. He said, 'Okay, that's the figure. I agree to it.' So we put it back together, and pretty soon, the show was back in business."

By the end of the war in 1945, it was clear that Kansas City's economy in the future would be less tied to agrarian activities. Between 1943 and 1944, 75 dairies closed in Johnson County, Kansas. Much of the land occupied by dairies and farms became the suburbs of the city. An example of this is the 1000-acre Wild Rose Farm which is now the site of Homestead Country Club in Prairie Village, Kansas.

A similar decline occurred in Jackson County: R. E. "Bud" Hertzog, D.V.M., the official veterinarian for the American Royal, has estimated that there were more than 250 dairies in 1956, and only two by 1998. "The urbanization of this area, the development of many of the farms, the high real estate prices have priced a lot of this land out of use for cattle production," Dr. Hertzog said, noting that the three largest breeders of Polled Herefords in Missouri used to be in the area, but are no longer there. A spring Dairy Show was held at the Royal from 1949 to 1951.

The American Royal building was still under lease in 1945. Fortunately, the weather in mid-October was in the '60s and '70s. A junior show was held. Price controls were still in effect, but were waived for the show, thanks to M. J. Flynn of the Wilson-Flynn Commission Co. "Mike Flynn, who was in the commission business here, was Captain Harry Truman's lieutenant in World War I. I went down to see Mike and told him our situation. He said, 'I'll call the president,'" Jay B. Dillingham recalled. President Truman took the call. "He said, 'Mike, what's your problem?' Mike told him, and the president said, 'I think I can handle that for you.' It wasn't fifteen minutes before we had a wire from the director of the Office of Price Administration, lifting that requirement on calves shown by FFA and 4-H members." The grand champion steer brought $2.25 per pound, the hog a record-setting $1.50 and the lamb $6.50, which was a $1.25 more than the world record at the time.

In the years after World War II, the name Utz would be prominent in the American Royal's horse show programs. Three brothers, Don, Jay and Sug Utz, were horse trainers. Many of Hazel Stone's horses, including Fleet's Country Boy, were trained by Don Utz. Jay Utz's most famous

Sug Utz and Contract's Commander (Photo by Jack Holvoet, courtesy of Sug Utz).

horse was the legendary The Lemon Drop Kid. Sug Utz went to work for Raytown banker Leroy Cox in 1945, and ran Mr. Cox's Blue Ridge Stables for more than 40 years. Sug Utz's first American Royal appearance was in 1946, and he has since showed for 52 consecutive years.

Among the winning horses he showed at the Royal were stallions Contract's Commander and Arletha's Stonewall's Masterpiece and three-gaited horses Bugle Ann, Mimi Genius, American Princess and Dream Doll. He won the Missouri-Kansas stake four times with Contact, whose dam was out of Lady Jane. He also won the Missouri-Kansas stake with a horse named Shamrock's Commander. After the horse show moved to Kemper Arena, he won the junior stake with Courageous Commander. Most recently, he had the second-place horse in the 1998 Tom Bass Missouri-Kansas Five-Gaited Stake. At 78, Mr. Utz was the oldest competitor. Mr. Utz is not merely a local favorite: Seven of his horses have been world champions, including Mimi Genius, American Princess and Dream Doll.

Sug Utz has taught many equitation riders, including Donna Hobbs Moore, Mariann Hobbs, Bussey Sofio and Suzanne Beu. His other pupils included Gail Watson Russell, Nancy Watson and Tina Hughes Sheary. Trainers who worked for him early in their careers include Mark Hulse, Nelson Green, Jack Cupp and Sonny Sutton as well as Ms. Moore.

The Bit and Spur Club, which was located upstairs at the south end of the arena, had been a popular meeting place for patrons, exhibitors and officials before the war, was restored in time for the 1946 show. The Kansas City, Kansas Service League took charge of decorating and furnishing the rooms. Committee members included Mrs. H. Darby Trotter (the former Betty Anne Jones), Mrs. Thomas Gibson (the former Harriet Darby, daughter of Harry Darby), Mrs. Robert M. Kissick, Jr., Mrs. Henry Holbrook, Mrs. Fletcher Steck, Mrs. Richard Granville and the league's president, Mrs. Joseph Stone McDowell. According to *The Independent*: "The theme for the decorations is Western, with an abundance of wagon wheels, saddles, spurs for curtain tiebacks and shiny boots to hold plants. The girls have painted the paneled walls with Western scenes, slip covered the furniture and dyed fabrics for drapes."

Horse show exhibitors of the postwar years fondly recall the old horse show arena. There was the long ramp. Olive Beaham (later Olive Wright; now Olive Lansburgh) recalled that once, when she and her horse were just starting to go up the ramp, the Budweiser Clydesdale hitch began coming down. Her horse, Dixie Belle, needed no encouragement to turn around. Ticket holders sat in boxes. The boxes, which contained wooden folding chairs with blue or gold covers, bore plaques inscribed with their sponsors' names. The walkabout allowed everyone to step up to the rail to see the action, and then stroll on to chat with friends. There was some mischief, too. Young exhibitors and grooms fought boredom during waiting periods by tying rubber spiders on strings and dropping them on the unsuspecting.

Horse show competitors often practiced in the middle of the night. One horse trainer recalled that, for many years, the week of the horse show was a time when he was lucky to get four hours of sleep per night: He and his competitors were up at 4 am to work the horses, showing them in the afternoons and evenings and rarely in bed before 1 am. Loula Long Combs was no exception: She was there at 4 am, with her hat on, and her Boston bulldogs at her side.

The American Royal horse show opened on Saturday, October 19, 1946 after a four-year hiatus. On the same day, the Civilian Production Administration announced that, effective immediately, all restrictions on women's clothing would cease: Long skirts would be available for the first time since 1942. America and the American Royal were moving back toward normalcy.

Allen Thompson had served as horse show manager from approximately 1924 until the beginning of World War II. After Mr. Thompson died in 1946, Charles W. Green assumed the management of the 1946 horse show. Edwin C. Eggert became manager in 1947.

The parade class was a very popular event, not least because the Royal was the site of the

Tina Sutton (now Sheary) and Let Me Entertain You at a Saddle & Sirloin Club horse show (Photo by Van Meter, courtesy of Wilma Hughes).

Buck Hinson and Cedar Creek Mr. Golden Heir (Photo © by Howard Schatzberg, used by permission, courtesy of Bill Paschall and Wilma Hughes).

world's championship. Audiences were enchanted by the high-stepping, slow-moving horses and by the costumes wore by their exhibitors. The 1946 $300 parade championship stake was won by Pancho and Mrs. L. E. Brown, with the red ribbon going to Mrs. Walter Gaugh and Sister Eileen, and L. Mondell Ahrenholz and Dice taking third place. Other participants who thrilled the crowd over the years included Danny Brakebill, Ted Bryant and Tina Sheary. In later years, H. W. "Buck" Hinson became the man to beat. Mr. Hinson, the founder of the Hen House Markets, always wore Western attire and cowboy boots. He rode Cedar Creek Mr.

Donna Moore (From photo © by H. Leon Sargent, used by permission).

Golden Heir to four World's Grand Championships in the Parade Division. Cedar Creek Mr. Golden Heir was sired by Ridgefield Heir, a champion stallion that Mr. Hinson bought from trainer Garland Bradshaw. Mr. Hinson's sterling-silver parade saddle and appointments, which he had owned for over forty years, were valued at $50,000 at the time of his death in December, 1997 at the age of 88. The parade class was dropped from the Royal's schedule in the mid-1990s.

Donna Hobbs (later Donna Moore) made the first of many appearances at the Royal in 1946. Miss Hobbs, then 15, finished sixth in the horsemanship championship for the junior division. She also took third place in the local five-gaited class. Ms. Moore's two daughters, Melinda and Melissa, are both accomplished equestrians. Their father is noted horse trainer Tom Moore.

Competition has always been a factor in the American Royal. In 1946, the most exciting competition occurred among the bidders rather than the exhibitors: "I'd always bought the grand champion steer — Williams Meat Company had all the way back to 1923," Eddie Williams later recalled. Representatives from several other companies decided to make it a little more challenging for Mr. Williams. "So they put a syndicate together, and I got word that they were going to outbid the last steer that was sold [at the previous show], which was $10,000 or $11,000. They figured $12,000 to $14,000 would do it. When I found that out, I knew I was going to have to spend $14,000 or $15,000," he said. The bidding grew intense: "I would have mortgaged anything but my wife," Mr. Williams recalled.

When the gavel came down, the steer, which had been raised in Iowa by sixteen-year-old Jack Hoffman, was sold to the Williams Meat Company for $44,375. According to Mr. Williams, "Some lady said, 'Lord have mercy! How's that kid going to pay that much money? Kearney Wornall turned around to her and said, 'I'm his banker. He's good for it.'" As Mr. Hoffman remembered it, he was stunned by the sale. "They told me I swallowed my gum," he said. The record held for more than fifty years. The sale record for the steer, T O Pride, is in the American Royal Museum.

Eddie Williams was frequently the highest bidder for grand champion steers, lambs and hogs. He attended his first American Royal in 1933 with his uncle, Lou Williams, who preceded him as president of the Williams Meat Co.

Eddie Williams made a practice of having three pairs of boots made from the hide of the grand champion steer: one for the owner, one for the American Royal president and "selfish me, a pair for myself." He later wrote, "I felt the owner of the steer would take his silver trophy around to livestock shows and fairs, but could wear the boots anyplace. Harry Darby told me . . . that he had better stop being president as he was running out of space to keep the boots." Senator Darby actually displayed the boots in his office, in front of a piece of rail fence with a saddle on it.

The purpose of the auctions always was to obtain the most money as possible for the education of the youngsters who raised the champions. In this, the auctioneers acted in tandem with sympathetic audience members. Auctioneers at the Royal have included M. R. Judy, who participated in the 1899 show, Arthur W. Thompson, Fred Reppert, who in the 1920s was probably the first auctioneer to hire a private plane to fly him to shows, Roy Johnson, Ray Simms, Hamilton James, Paul Good, Jack Halsey, Gene Watson, George Morse, Jewett Fulkerson, Charlie Corkle and Stanley Stout.

Mr. Williams often opened the bidding by holding up one finger — thus indicating a starting bid of one dollar. The auctioneers, including Roy Johnson and, later, Ray Sims and Glen Bratcher, then egged audience members on to bring the bids higher. Eager buyers of grand champions over the years have included restauranteur Jud Putsch of Putsch's 210, who was Eddie Williams' brother-in-law, restauranteurs James and Don Lee of the Savoy Grill, the Golden Ox restaurant, chain store magnate J. C. Penney, Ford dealer Berl Berry and Tom Williams of Guaranteed Foods. According to writer William Franklin, "J. C. Penney once sent several of his store managers to an American Royal auction to up the bids on a lamb owned by a young Oklahoma boy dying of bone cancer." Mr. Berry was a colorful character who once ran unsuccessfully against H. Roe Bartle for mayor. After purchasing a grand champion steer named Dorothea's Pride, he decided to display the animal at his used car lot at Armour and Main Streets. The animal was reported stolen the following day, and soon turned up in its preferred habitat: a pasture north of town.

The American Royal experience for the sellers of the champions is bittersweet: Youngsters find parting with a prized lamb, calf or steer to be a heartrending experience, in spite of the awards and the money. Nearly seventy years after the event, John Riffle wrote:

> Mr. Lou Williams purchased an Angus steer from me at the American Royal for the
> West End Cafe which was across the street from the Livestock Exchange Building and

John Riffle and his steer Jimmy (Photo by Guy E. Smith, courtesy of Col. and Mrs. John Riffle).

Dr. Don L. Good with FFA member David Deason and grand champion FFA steer, 1964 (Photo by William L. Glover, courtesy of the Western Historical Manuscript Collection-Kansas City).

where many of us ate while attending the Royal. I led the steer down from the sale ring in the arena and tied the steer to a lamp post in front of the West End Cafe, and I have to confess: I cried like a baby. I hated to give that steer up. We had been buddies from one November to the next. But after you have raised several steers and sold them, you realize that this is something that happens.

Some decide to give up showing market animals. Others, such as Col. Riffle, accept that this is a part of agribusiness and resolve to raise next year's champion.

The 1946 American Royal was sold out almost every night. Andy Paterson described it at the time as "Kansas City's richest Royal." At the end of the week, Roy Roberts, managing editor of the *Kansas City Star* wrote an article in which he enthused:

My faith in my country's future has been renewed. The American Royal has done it.

It has not been the fine horses, the great bulls and steers and calves. Nor the huge hogs, so large and so fat you wonder that they can stand the heat and the travel — what luscious pork chops in the making! All that and more has been paraded before the Middle West on a scale and a grandeur that marks this revival of the American Royal as the greatest ever — the forerunner of still greater to come.

The real soul of the American Royal, however, is not all this, which has been so proudly paraded and exhibited before tens of thousands. The soul of the American Royal has been the crowds, especially the thousands upon thousands of farm boys.

Don L. Good was one of those farm boys. He was the high man in the livestock judging contest in 1946 as an Ohio State University student. "We stayed at the Continental Hotel. Of course, we were poor, and we had to put three and four people in a room so that we didn't spend much money. Nonetheless, we thought it was quite an experience." The team observed 12 classes of animals, including cattle, horses, swine and sheep, and judged eight. Herman Purdy coached the team. At that time, he was simply the new coach, hired after the war. Later, he made a name for himself as a judge, both in competitions in the United States and in Canada, Argentina, Australia, Brazil, Uraguay, Costa Rica and Scotland. "Herman Purdy probably judged more beef cattle shows than any other man who has ever lived," said Dr. Good.

Dr. Good's adviser at Ohio State University was Don Kays, who was the head of the animal husbandry department and who served as a cattle judge at the American Royal. Dr. Kays was a great friend of Eddie Trowbridge, the dean of agriculture at the University of Missouri, and the two of them often judged saddle horse and hunter/jumper competitions. Dean Trowbridge also judged livestock and served as a director of the American Royal. According to writer Dale F. Runnion, Dean Trowbridge "was a keen livestock judge of all classes of livestock, excelling as an authority and judge of all classes of horses." The Trowbridge Center at the University of Missouri at Columbia is named for Dean Trowbridge.

There have been many highly qualified, highly respected livestock judges at the Royal. Of Warren Lale Blizzard, dean of agriculture at Oklahoma A & M College (now Oklahoma State University) who was a cattle judge, Mr. Runnion wrote, "He judged practically every major livestock show in North America. His wise judgment and counsel on quality and bloodlines were widely sought." Beef cattle judges in the Royal's early years included Glenn Bratcher, who later served as executive vice president of the American Angus Association, Arthur McArthur, Dr. A. E. "Al" Darlow, Dean W. L. Carlisle, Dean H. H. Kildee, Dr. Hilton M. Briggs, Dr. Al Dyer, John H. Skinner, Dr. A. D. Weber, Phineas S. Shearer of Iowa, Clinton K., John R. and James G. Tomson, Dean C. F. Curtiss, Jack Turner, F. W. Harding, Otto V. Battles, John Burns, A. "Wood" Harris, Robert W. Lazear, Herbert Chandler, William H. Pew, J. Milt Tudor, Otto G. Nobis, J. Garrett Tolan, and Jack Van Natta as well as Ralph Reynolds, Cal Kinzer and Kenneth McGregor.

Dr. Good later coached winning teams from Kansas State University and served as a beef cattle judge for many of the Royal's shows and for many other leading fairs and stock shows.

Dr. Good fondly remembered the livestock judging awards breakfasts that were hosted by Thomas Wilson of Wilson & Co. "There was a lot of fellowship there, a lot of friends made," Dr. Good said. The Chamber of Commerce used to give canes to attendees at the breakfast. Mr. Wilson owned Edellyn Farms in Illinois, where he raised Shorthorns. In later years, his son, Edward, hosted the breakfasts.

"I think that you'll find that many of the alumni of the American Royal 4-H judging contests and college livestock judging contests will be in leadership roles throughout the entire livestock and meat industry as a result of their experiences, and the stimulation and inspiration which they received at the American Royal," he said. "Livestock judging teaches memory. It teaches careful analysis and careful observation of animals. Above all, it teaches communication skills." Dr. Good was head of the animal science department at Kansas State University from 1967 until his retirement in 1987.

Two notable retirements were recorded in 1946. W. A. Cochel, agricultural adviser to the *Weekly Kansas City Star*, farmer and breeder of Shorthorn cattle, was awarded a citation by the FFA at their convention. "His efforts have inspired members to increase their achievements and improve their way of life on the farm. May God continue to bless him as he retires to his farm, Roanridge," it read in part. Mr. Cochel had been the presenter of the Star Farmer of the Year Award since its creation in 1928. Easter Serenade, a mare belonging to Temple Stephens of Moberly, Missouri, was retired in front of a sellout crowd. The mare had won her first American Royal stake in 1938, and had gone on to receive several blue ribbons in both the fine harness and the five-gaited stakes. Nowhere near retiring, Loula Long Combs was presented with a silver trophy in honor of her golden anniversary in the show ring.

As in previous decades, Mrs. Combs was exhibiting prize-winning horses. In October, 1947, the *Kansas City Times* ran a photo of her bay gelding Radiation in his stall with his two companions —

a pair of bulldogs named Butts and Jiggs. Radiation had been the 1946 Champion Hackney Pony.

The American Royal began operating as a tax-exempt organization in 1948. The fact that the Royal was granted this exemption shows that Jackson County judges felt that the show provided education and worthwhile entertainment; in those days, the judges had the power to grant or deny such requests.

A letter from the White House to Harry Darby:

April 1, 1948

My dear Harry:

It makes me very happy to have a reminder through your letter of March eighteenth that my old friends have not forgotten my interest in good livestock. I have much pleasure in accepting membership (honorary) on the Board of Governors of the American Royal Live Stock and Horse Show. I am made doubly happy by the fact that my election was unanimous.

In thanking the members for thus honoring me please tell them that their action has brought back a flood of memories of old associations. My father and I were familiar with the founding of the association and were early and enthusiastic exhibitors.

With personal good wishes,

Always sincerely,

(Sgd) Harry S. Truman

Two stables which opened in 1948 would have long-term ties to the American Royal. Jane and Lee Fahey moved to Kansas City and opened the Fahey Farm and Stable. Mr. Fahey had previously worked as a trainer for George Godfrey Moore in Topeka and for Carl Goetz in St. Joseph. Mrs. Fahey, the former Jane Fairchild, had operated Lake Contrary Stables in St. Joseph. Horses from the Fahey Farm included Anacacho Empire, a stallion Mrs. Fahey showed successfully in the parade division for many years. The Faheys were inducted into the UPHA Hall of Fame in 1990.

Virginia and Don Hulse established the Don Hulse Stable, now the Hulse Stables, in Center, Missouri. The Hulses purchased Champagne Fizz when he was a yearling. Champagne Fizz's progeny includes Champagne's Mr. Perfection, winner of the UPHA classic five-gaited grand championship, and Heir to Champagne, as well as Champagne Heiress and Champagne Debutante, both of which Mr. Hulse showed at the Royal.

The Independent issued its own judgment on the 1948 show:

Getting parked this year was as bad as ever, with husbands shouting at parking attendants who wave them farther away from where they want to go, and the wives teeter perilously on high heels as they make their way over the cobblestone streets of the West bottoms to the entrance. Once inside the building such troubles are forgotten with the old thrill of seeing the gay bunting and emblems on the stalls, representing stables from all over the country. The men always must see the cattle on exhibition and the aforesaid ladies in high heels begin to wince a bit as they wander along the vast underneath of the Royal. In the boxes, however, the ladies come into their own, with their hats a strong competition to events in the show ring. Sneezes and handkerchiefs do not indicate an unhealthy audience, but are results of tanbark thrown in the air by a few dozen hooves on a fast track . . . Many "raunchy cowhands" led their quarter horses around the ring, and in among them were Barbara and Bob Sutherland, in levis and crushed hats. During the judging of the hunters, Dana Durand turned in a

The 1940s

Aerial view of the American Royal and the Sutherland Lumber Company (Photo courtesy of the American Royal Association).

sterling performance, leading in one of the mounts, ridden by his wife, to be judged. . . Carol did well, her steeds winning two red ribbons.

Robert Q. "Bob" Sutherland was the author of a book, *The Quarter Horse As I See Him.* He became horse show chairman in 1961, succeeding Dallas Alderman, and served through 1974. The American Royal Museum showcases many items which belonged to him, including his chaps. His family owned Sutherland Lumber Company, which was located across the street from the American Royal. Mr. Sutherland was from a family of five children: a daughter, Donna, and four sons, Dwight, Herman, John and Robert. His brother, Dwight Sutherland, served as president of the Royal in 1983-1984 and is the honorary co-chairman, with R. Crosby Kemper, Jr., of the 1999 American Royal Centennial.

Carol Hagerman Durand was selected for the 1952 Olympics; when it was decided that women would not be allowed to compete against men, she allowed another rider to take the horse she rode, Miss Budweiser, which was owned by August A. Busch, Jr. Mrs. Durand was killed in a riding accident in 1970. The Carol H. Durand Challenge Trophy is awarded in memory of her. Her husband was president of the Saddle & Sirloin Club in 1961-62 and served for several years as a horse show steward at the American Royal.

R. J. Kinzer (Photo from *The Story of the Herefords* by Alvin H. Sanders).

The 1948 American Royal included a special Hereford show, known as the "R.J. Hereford Royal" in honor of R.J. Kinzer, who had retired as secretary of the American Hereford Association in 1944 after 33 years of service. Mr. Kinzer had also served as general manager of the American Royal for one year, 1915. The Golden Anniversary Hereford Show in 1932 was his idea, and to ensure its success, he arranged to offer $50,000 in prize money. The 1948 event attracted 77 breeders from 24 states. The *Hereford Journal* noted that "the high quality of the animals competing for the $75,000 in cash prizes was the kind that brought keen satisfaction to the man whose work over the years had contributed to such a great extent to Hereford promotion and advancement." Mr. Kinzer died in 1952 while inspecting a breeder's herd in Florida. "You're all right, old fellow," he said, patting a bull on the back. As he turned to walk away, he suffered a fatal cerebral hemorrhage.

The spring of 1949 saw the first annual dairy show at the American Royal and, more important, the first American Royal Rodeo. The rodeo that year was memorable: It featured the award-

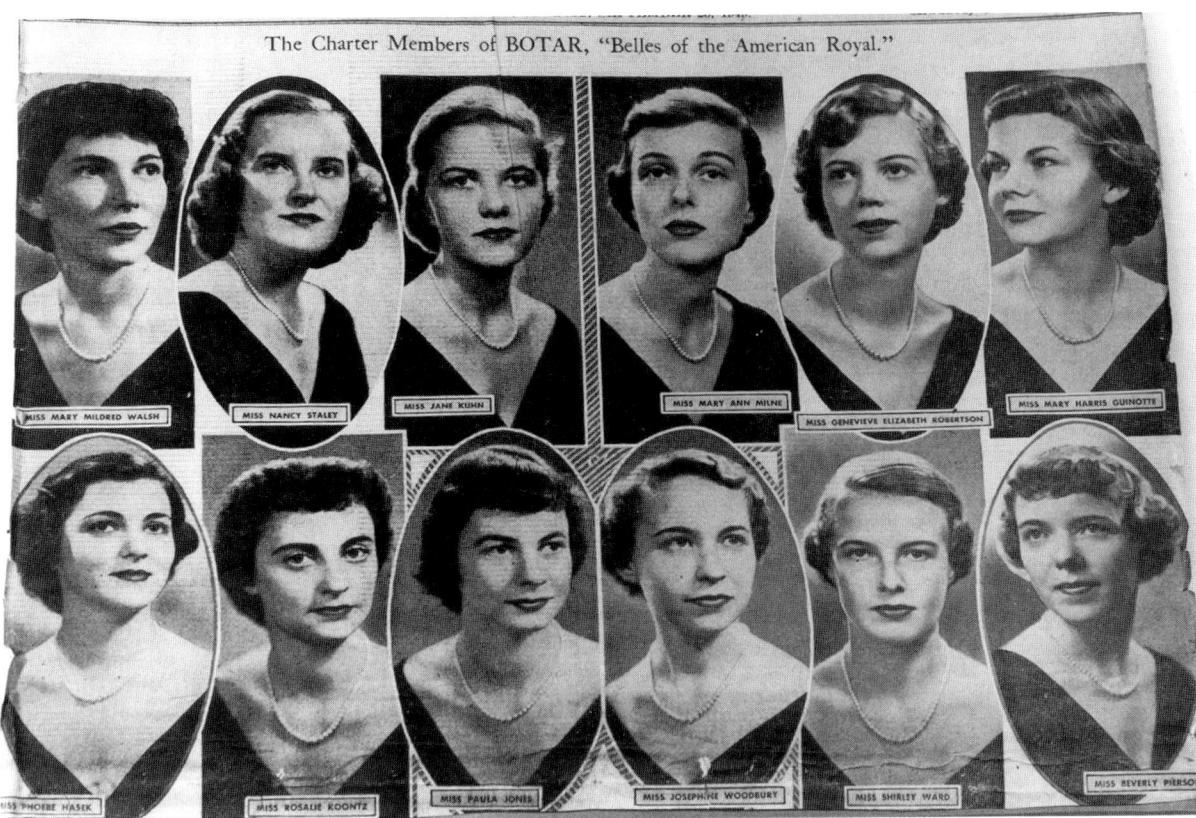

The Charter Members of BOTAR, "Belles of the American Royal."

MISS MARY MILDRED WALSH · MISS NANCY STALEY · MISS JANE KUHN · MISS MARY ANN MILNE · MISS GENEVIEVE ELIZABETH ROBERTSON · MISS MARY HARRIS GUINOTTE

MISS PHOEBE HASEK · MISS ROSALIE KOONTZ · MISS PAULA JONES · MISS JOSEPHINE WOODBURY · MISS SHIRLEY WARD · MISS BEVERLY PIERSON

The BOTAR Class of 1949 (Photo courtesy of the BOTAR Organization).

winning cowboy Jim Shoulders, world's champion steer wrestler Homer Simpson and legendary bullrider Freckles Brown. The dairy shows, which ran for three days per year, were promoted by the Mid-America Dairymens Association, which was then headquartered in Kansas City. The dairy show and the rodeo were discontinued after the spring show of 1951.

Bud Hertzog was a boy at the time of the dairy shows, which were part of his early exposure to the American Royal. "I grew up on a dairy farm. I showed Angus cattle as a 4-H member and a FFA member. While I was in college, I was a member of the University of Missouri livestock judging team in the livestock judging contests at the American Royal. And then I ended up going back and treating animals there," he said. Dr. Hertzog has been the American Royal's veterinarian for more than a quarter of a century, and also oversees a internship program at the Royal for veterinary students from the University of Missouri and Kansas State University.

In 1949, the Belles of the American Royal, or BOTARs, were founded by 10 supporters of the American Royal. Their goal was to foster interest in the American Royal among young adults. That year, 12 young women, the first BOTARs, were presented, with their escorts, at the Coronation Ball. *Life* magazine covered the gala. The Coronation Ball was held in honor of the American Royal Queen, and the BOTARs originally were meant to serve as her court. Only once was a BOTAR chosen as American Royal Queen: Nancy Moore (later Thornton) in 1958. The BOTARs were selected by a committee known as the "Secret Six." The silhouettes of these three mystery women and three mystery men graced the program until 1959, when the BOTAR Organization was formed. The BOTARs appeared at the Coronation Ball for the American Royal Queen until 1970, when the first BOTAR Ball, a fundraiser for the Royal, was held at the Muehlebach Hotel. BOTARs, as members of the BOTAR Organization, have given over $1,250,000 to the American Royal; as individuals, they have given even more.

The BOTAR Class of 1953 (Photo courtesy of the BOTAR Organization).

BOTARs are active in all aspects of the American Royal. These career women and volunteer leaders head American Royal committees and serve as directors of the Royal, arrange escorts for the student ambassadors, and, individually and as an organization, buy animals at auction. BOTARs lead museum tours for schoolchildren during the American Royal. The BOTAR Organization funds scholarships to the first, second and third place high point individuals in the Intercollegiate Livestock Judging Contest. On BOTAR Night at the horse show, amateur exhibitors compete in a three-gaited class sponsored by the organization. The BOTAR organization also sponsors graduate credit courses for area schoolteachers. These courses focus on bringing agriculture into school curriculums and are held at the American Royal. The popular Children's Horse Show is a BOTAR project. BOTARs provide leadership, vision and research acumen to the education committee, and played an important role in the creation of the American Royal museum. The BOTAR commitment to education, scholarships and volunteerism has helped to make the American Royal the grand event that it is today. BOTARs devote untold hours without fanfare or desire for recognition, but simply for love of the BOTAR Organization and the American Royal.

For the ball, BOTARs wear matching gowns. The color and the style of the dresses changes annually. Inevitably, the BOTARs carry the white ostrich plume fans that are their trademark.

The first live television broadcast in Kansas City occurred at the American Royal in 1949. Senator Darby and Jay Dillingham appeared on WDAF-TV Channel 4 with Roy Roberts of the Kansas City Star, which owned WDAF-TV, and reporter Randall Jessee. Senator Darby wore cowboy attire for the show, and Jay Dillingham later recalled that people who saw the broadcast said the picture came through clearly.

The BOTAR Class of 1980 (Photo by Strauss-Peyton, courtesy of the BOTAR Organization).

The 1950s

"Mom, when I get out of this, sister and I are going to put Beckham county on the map. We're going to win grand championships all over the American Royal," Benny Hopkins wrote in 1945. The 19-year-old soldier, who had won prizes for steers, Poland China and Berkshire hogs and Shropshire fat lambs, had already seen his final American Royal: That letter was the last one he sent home before he was killed in a bomber crash in Europe.

Minnie Marie Hopkins receiving a calf from John Vanier (Photo by permission from Joyce Hale from *Fifity Years of Kansas Ranching: CK Ranch* by Bud Snidow).

Five years later, his sister, Minnie Marie Hopkins, brought a Hereford steer named Misty to the show. Miss Hopkins, 15 at the time, had bad news while at the Royal: Her other steer, George Bob, died in Oklahoma after licking insecticide powder. Misty placed seventh in a field of 54. "When I sell Misty at auction Friday, I'll bank the money until I can find two more good Hereford calves to feed and get ready for the Royal," she said. "Benny and I are going to get a grand champion before I stop." After the sale, Miss Hopkins was called back to the sale area by Jack Turner, secretary of the American Hereford Association. John Vanier, owner of CK Ranch in Brookville, Kansas, presented Miss Hopkins with a six-week-old purebred Hereford calf. He told her, "This is CK's way of showing that we admire your courage and we hope your dreams of having a champion will come true with this calf." The Vanier family has continued its tradition of generosity at the American Royal for several generations: Hale Arena is named for John Vanier's daughter and son-in-law, Joyce and Joe Hale, who have been enthusiastic supporters of the Royal. The Hales' daughter, Dana Nelson, is co-chairman of the American Royal Centennial committee with Marianne Kilroy.

Lon Cox came to the American Royal as assistant horse show manager in 1950. He later traded jobs with Edwin C. Eggert and became horse show manager, a job which he held until 1968. Shirley Parkinson described Mr. Cox as "an outstanding manager, very well-liked." Legend has it that on at least one occasion it was Mr. Cox's eloquence that allowed the show to go on: Several grooms had been arrested by the police for playing craps backstage at the Royal. The remaining grooms threatened to walk out. Mr. Cox interceded on behalf of the Royal, and the police agreed to drop the charges.

The 1950s

Lon Cox, horse show manager, circa 1950s (**left**) (from a photo © by H. Leon Sargent, used by permission, courtesy of the Western Historical Manuscript Collection-Kansas City).

Floods damaged the West Bottoms in 1903 and again in 1908. But the Flood of 1951 hit hardest. "This morning, when workers arrived the yards were dry and the pens cleared of cattle. However, the view from the west windows of the upper floors [of the Live Stock Exchange Building] showed water covering the packing house area across the Kaw," the *Kansas City Star* noted on Friday, July 13, 1951, recording that:

Long before floodwaters surged over the dike and the Rock Island and stockyards bridges, the Live Stock Exchange building had been evacuated on order of Jay Dillingham ... Later the Turkey Creek pumping station was shut off and vans and trucks were removed by volunteer drivers as water rose to a 15-foot depth in the American Royal building... Jay B. Dillingham, president of the Kansas City Stock Yards company, and Cliff J. Kaney, president of the Kansas City Live Stock Exchange, were exerting every effort to get the workers in the building out of the danger zone.

Only first floor offices of the Exchange building were finding it necessary to move supplies and records to upper floors. Occupants of upper floors merely turned off the lights, closed the doors, and walked out.

The evacuation was quiet. Noon was generally considered the deadline. Most of the workers leaving the building were more interested in obtaining a closer look at the flood threat. Many visited the dikes before leaving the area.

In 1991, Donald Ornduff remembered walking back from lunch that day:

Our office was in the building at 10th and Wyandotte — and there were a lot of people coming down the hill from the west. I said, "What in the world is the attraction?" to somebody I met. He said, "There's a heck of a flood." Instead of going to work, I just kept walking up the hill. I looked down there and couldn't believe what I saw — automobiles floating down, cattle floating down, hogs floating down. Water was up to

Clipper Lady, a Shorthorn heifer belonging to Geo. Struve & Sons of Manning, Iowa, 1950. According to Bud Snidow, photographer Guy E. Smith was known and respected for his ability to retouch pictures of animals. Because the desired look at the time was for the animal to be compact, Clipper Lady's legs were buried in straw, and the photograph was shot from above. Mr. Smith then corrected the line of the heifer's back so that it appeared straighter and created a lustrous tail for the animal. (Photo by Guy E. Smith, courtesy of the Western Historical Manuscript Collection-Kansas City, American Royal Collection).

the top of freight cars and loading docks and warehouses down in the bottoms and then it began to go over the levee which was designed to protect the bottoms from the Missouri River. Instead, the water was running in reverse. All the water in the stockyards was from the Kaw.

Jay Dillingham stayed at the Livestock Exchange Building until 9 pm that night. With Mr. Dillingham was Bob McCreight, then vice president of the Kansas City Stockyards Company. "We made sure everybody had gotten out safely and then we went down the third floor fire escape to a motorboat," Mr. Dillingham later recalled. "Hogs were walking around on the fifth and sixth floors."

Damage to the American Royal from the 1951 flood would exceed $85,000. Everyone involved was determined to see the show open on time. The Kansas City Stock Yards Company restored the building. In order to finish quickly, three groups of painters were brought in to spray-paint the walls of the American Royal. They worked virtually on top of each other; the doors of the building were kept locked, for fear that a city inspector would arrive and, on seeing the disregard for safety precautions, shut the project down. The American Royal opened as scheduled — to a full house.

Auction at the Royal. This photo is believed to have been taken after the 1951 flood. The auction was held in a barn usually used as a garage. Note the cars in the background (Photo by Randozzo & Morrison, courtesy of Marie Schubert).

Roger Shores at the American Royal, 1954 (Photo by Launspach, courtesy of Roger and Melissa Shores).

This was a message to the world that Kansas City was back in business.

One person who was apparently undaunted by the flood was Loula Long Combs. At the American Royal Horse Show on October 15, she collected three blue ribbons: first place in the Hackney Pony Division, Ladies Single with her bay Hackney Possession, in the Heavy Harness Open Pair class with her brown geldings Citation and Constellation and in the Heavy Harness Division with her brown stallion Vibration.

Sue Noll first became involved in American Royal activities when she worked on the 1951 parade. She was then secretary to Dick Challinor, the Aviation Commissioner for the Chamber of Commerce, who was also secretary of the American Royal Parade Committee. "When he left the Chamber, I just inherited the job," she recalled. She went to work for the Royal in 1970 as a member of the horse show staff. In 1976, she became secretary to Laurence Pressly, who served as general manager of the Royal from 1977 until 1985. When she retired after the 1985 show, she had organized 34 parades.

Roger Shores was a lad of seven when he first rode in competition at the American Royal in 1952. He credits Loula Long Combs with the beginning of his family's involvement with horses: "Our farm was about a quarter of a mile from Longview. We didn't know anything about horses, but because of the proximity of her farm, we became interested in the horse business. We bought many of our horses there, and our trainer came from Longview." Mr. Shores treasures a copy of her book, *My Revelation*, which Mrs. Combs inscribed and gave to him:

> With kindest wishes for Roger and hoping he will enjoy his ponies as much as I love mine. From your sincere admirer.

Mr. Sandman, shown by Charlie Bishop (Photo courtesy of Roger and Melissa Shores).

Loula Long Combs, January 1955

Mrs. Comb's hopes were well-placed: Mr. Shores has served as the horse show chairman since 1992.

Mr. Shores' parents, Dorothy and Ralph Shores, owned many Hackney ponies. Mrs. Shores liked to name their horses after popular songs: My Little Margie, Peg O' My Heart, Autumn Leaves and Mr. Sandman. The most successful of these was Mr. Sandman, which the Shores co-owned with Warren Newcomer. Mr. Sandman was trained by Charlie Bishop and won his fifth harness pony stake in 1962. The

Shores also owned Miss Patricia, which won the American Royal parade pony class in 1953, 1954 and 1955.

"Ladies and Gentlemen, history has made here tonight..." Ralph Boomer, the Royal's announcer said before he was drowned out by the audience's applause. The Replica, a chestnut gelding owned by Mary Jane McGrath, won the five-gaited championship for the third time on October 20, 1951, retiring the *Kansas City Star*'s Challenge Trophy. Art Simmons was The Replica's rider that night. Marty Mueller had shown him in 1950, and Lee Roby started the winning streak in 1949. Other exhibitors who had attained two (but not three) since the trophy was first offered in 1933 included Carnation Stables, Leisure Hour Stables, and Temple Stephens.

Mary Gaylord McClean (Photo © by Tod Macklin, used by permission).

Lynn Weatherman, now senior editor at the American Saddlebred Horse Association, was an Iowa schoolboy when he began attending the American Royal in the early 1950s. "It was the highlight of the year," he recalled. "My parents were both schoolteachers, and they would set aside part of that week for vacation and drive down." Frances Dodge van Lennet of Dodge Stables was a prominent exhibitor in those days. Mr. Weatherman remembered Lee Roby of Green Hill Farms as "probably the best horse trainer there was." Mary Gaylord McClean was among the riders whom Mr. Roby instructed. Mr. Weatherman saw Winged Commander, described by announcer Ralph Boomer as "one of the greatest show horses of all time," win the five-gaited stake in 1952, but he described Lady Corrigan, a bay mare owned by Josephine Abercrombie and trained by Garland Bradshaw, as "the most thrilling five-gaited horse I've ever seen in my life . . . Every step, it looked like she was going to blow up, and she never did. You'd be on the edge of your seat. She was beating everything in the class, but you were afraid she was going to jump out of her skin and beat herself." Lady Corrigan won the five-gaited championship in 1954.

There was tremendous excitement at the walking horse championship in 1952, as Lynn Weatherman recalled:

> There was a young horse from Iowa named Midnight Secret. The reigning world's champion was there. His name was Talk of the Town. They had three workouts. In the last workout, the two horses were by themselves — and yet they ran into each other. The class went to Talk of the Town. I thought there was going to be a riot . . . People were stomping their feet. I thought that old building might collapse. They were shaking it and rocking it. They booed the judges, they threw their programs, they threw coins. When it was announced that Midnight's Secret was second, I bet they cheered for fifteen minutes. It stopped the horse show.

The crowd believed that the judges had favored Talk of the Town because he was the world champion, but the situation was more complicated. Mr. Weatherman learned later that the judges were aware that Midnight's Secret had been sored. Soring occurs when a chemical or sharp object is applied to the pastern (the area between the hoof and the ankle) of a horse's front legs. The horse then puts more weight of its hind legs, and its stride appears more animated. Soring is now illegal under federal law; in the late 1960s, however, the practice became so widespread that the American Royal ended its walking horse championships.

John B. Gage served as president of the American Royal in 1953, succeeding Harry Darby. Mr. Gage had been mayor of Kansas City in the 1940s. He was a co-founder of the Saddle & Sirloin Club. His son John C. Gage, who served as American Royal president in 1972, recalled that his father owned farms where he raised registered Shorthorns. John B. Gage II, the son of John C. Gage, currently serves as chairman of the parade committee and as a director of the Royal.

According to Eddie Williams, Mayor Gage was the first to host a breakfast for cattle owners and herdsman prior to the steer auction. The event proved popular, and was continued by Bernice Van Voorst Hanback (whose family business was HyPower Chili) and later by Betty Calvin and Mr. Williams. Everyone would dress up and meet at the Golden Ox at 5:30 or 6 am for a hearty chuckwagon breakfast of steak and eggs, whiskey sours and Bloody Marys. From there, they would proceed to the sale barn.

In 1953, with the Korean War just ended, President Dwight Eisenhower came to Kansas City for three days, in which he dedicated the new Hereford Association Building at 715 West 11th Street, addressed the FFA at Municipal Auditorium and spent a great deal of time at the American Royal.

Security for government officials was more relaxed in those days. The night the president arrived, the arena was crowded. All the lights were off, except for the spotlights on the Herefords. Each animal was led by two handlers, who wore white coats. Above the heads of the audience were the vast windows which encircled the arena. The leader of the Secret Service group accompanying President Eisenhower took one look at the scene and asked, "Who will vouch for all these people?" Senator Darby pointed to Jay B. Dillingham. "He will," he said. This was apparently enough; The president attended the show.

President Eisenhower's visit to the American Royal in 1953 was recorded by *Horse World* magazine:

> There was a brief appearance of some 40 riders of the Saddle and Sirloin Club, introduction of the Royal Queen and her two Princesses, then a parade of 200 head of prize 4-H and FFA stock . . . The ladies parade class showed before President Eisenhower. When it was lined up he leaned over to [Harry] Darby and speculated on who would be the number one and number two horses. Ike's judgment was verified a second later when announcer Ralph Boomer named the same winners he had picked: Anacacho Empire, owned by Lamer Hotels, Jane Fahey up for first and Miller-Pontiac's Peavine's Golden Major for reserve.

Ezra Taft Benson, the United States Secretary of Agriculture, and Oveta Culp Hobby, the United States Secretary of Health, Education and Welfare, accompanied President Eisenhower. The president's visit was yet another proof of Harry Darby's ability to lure his friends to the American Royal. The president was himself a cattleman — he raised Angus on his farm in Gettysburg, Pennsylvania. When he showed his herd to Nikita Khrushchev, the premier of the U.S.S.R., President Eisenhower said that he was more proud of his cattle than of his medals.

Children in the audience in 1953 were thrilled by the presence of Duncan Renaldo, known to television audiences as "the Cisco Kid" and his sidekick, Poncho, played by Leo Carrillo. After the show, the actors stood by the window and greeted their fans.

Between the horse show classes, there were exhibition acts. The Rodeo Kids, a group of trick-riding youngsters trained in Lee's Summit, Missouri, have been audience pleasers since 1939. One of the popular acts of the 1950s was a grey mare named Sophisticated Lady, who was trained by Austin Smith. Sophisticated Lady did high school dressage exhibitions in harness. "This mare was so bril-

President Eisenhower at the FFA's 25ᵗʰ annual national convention (Photo courtesy of the American Royal Association).

liant, she inevitably got a standing ovation," Mr. Weatherman recalled. In more recent days, crowd-pleasing specialties acts would include Mexican charro Gerardo "Jerry" Diaz, John Payne, known as "the one-armed bandit," who rides horses and drives cattle, and Tommy Lucia, who has a monkey that rides a dog while the dog herds sheep.

Miss Budweiser, a gray mare owned by August A. Busch, Jr. excelled in the jumper events of the 1953 American Royal. The mare was already an Olympian, having participated in the games in Helsinki. The Kansas City Star recorded her performance in the modified Olympic stake at the Royal in 1954:

> In the Olympic events, the crowd cheered while Miss Budweiser, which reportedly cost $20,000, sailed over six barriers of more than four feet, made the difficult Liverpool jump over a 4-foot hedge and across a 6-foot-wide moat and then leaped between two flares to win.

It was not Miss Budweiser's first victory that day; during the matinee horse show, she won the sky-scraper class. Her rider was Robert Egan.

L. Russell Kelce served as president of the American Royal in 1954 and 1955. Mr. Kelce was

The Rodeo Kids (Photo by Leonard Hunting, courtesy of the Western Historical Manuscripts Collection-Kansas City, American Royal Collection).

a Shorthorn breeder who kept cattle at his 1,000-acre Grandview, Missouri farm and his 3,500-acre farm near Hume, Missouri. He was president of the Sinclair Coal Company.

Bud Snidow first came to the Royal in 1955, the year he moved to Kansas City. He recalls that cattle exhibitors in those years were required to sign in by Saturday morning before the show. The cattle remained on the grounds until the following Saturday night. Herdsmen had to stay with the cattle throughout the week. "We used to have 'tie-outs,'" he said. "Instead of keeping the animals tied up inside the building, the show provided a place on the outside. The cattle could be led outside, tied up and kept out overnight. Early in the morning, before the cattle were brought back into the barn, the fellows would renew the straw and the bedding." Many herdsmen slept in the straw.

The year 1955 was big for entertainment at the Royal. The "Today" show with Dave Garroway and chimpanzee J. Fred Muggs (who reportedly earned $975 per week — not bad money for a primate) was broadcast from the American Royal Building. In 1949, the Royal had been the site of the first local television broadcast in Kansas City; six years later, the "Today" show staged one of the first national broadcasts from Kansas City at the Royal. Eddie Fisher made an impromptu appearance at the FFA Convention. Mr. Fisher sang briefly, and Debbie Reynolds, then his wife, delighted the group by slipping off her mink coat and donning an FFA blue jacket. The couple was in town for the Coronation Ball.

Lee Roby, left, and Lloyd Teater, right (From photo © by H. Leon Sargent, used by permission).

The 1956 horse show awards sheets read like a Who's Who of important exhibitors, trainers and horses of the era. Jay Utz won the fine harness class with The Lemon Drop Kid, a chestnut gelding owned by Sunnyslope Farm. Shirley Parkinson recalled an incident with the horse that made the audience laugh: "He got loose one night and came down the ramp without a driver, hooked to a fine harness buggy." The Lemon Drop Kid appeared on the cover of *Sports Illustrated* in November, 1957 — a rare distinction for a harness horse.

In the five-gaited championship stake, the winner was Red Gold, a chestnut gelding owned by Miss Susan Richtmyre and ridden by Lloyd Teater. Second place went to Pinetree Genius, a chestnut gelding belonging to Judy Marks Pinetree Stable, with Chris Reardon aboard. Sally McClure (now Sally McClure Jackson) rode her chestnut mare, Golden Butterfly Again, to take third. Jean McLean Davis and Twilight Walk, her chestnut mare, were fourth.

The three-gaited championship stake was won by Bugle Ann,

a chestnut mare owned by Mr. and Mrs. E. H. Green and shown by Art Simmons. Sunshine Carol, a chestnut mare owned by Miss Molly Moody and shown by Garland Bradshaw was second. Mimi Genius, a chestnut mare belonging to Earl B. Noel and Wyman Bennett and shown by Sug Utz, took third. (Mr. Noel and Mr. Bennett had paid $80 for Mimi Genius at a dispersal sale. Years later she would sell for $25,000.)

Other trainers at the 1956 show included Dorothy Macleod, Jane Fahey, Rex Parkinson, Welch Greenwell, Charles Judd, Don Utz, Chat Nichols, Chris Reardon and Martin Mueller. Nathalie M. Nafziger, daughter of the Roy A. Nafzigers, Patsy Mackey, daughter of the Joe Mackeys, and Pie Coons were among the locally prominent exhibitors present.

Loula Long Comb's bay geldings, Vocation and Avocation, driven by Dave Smith, placed third in the Hackney pony — open pair — under 14.2 hands class; first prize went to a pair from Mrs. Wm. P. Roth's Why Worry Farm. Mrs. A. C. Thompson, another longtime competitor, drove a pair herself, but failed to place. There were eight horses entered in the heavy harness championship stakes. Three were owned by Loula Long Combs, and they placed first, second and fourth respectively: Dave Smith exhibited the blue ribbon winner, and Johnny Haffey's horse took the red.

Jay Utz and the Lemon Drop Kid (Photo © by H. Leon Sargent, used by permission).

Art Simmons, 1958 (Photo courtesy of the Western Historical Manuscripts Collection-Kansas City).

Several exhibitors at the 1956 show would later make headlines outside the horse world. Joan Robinson (later Joan Robinson Hill) was known for dying her hair gray to match her mare, Beloved Belinda, with whom she won the amateur stake in the five-gaited division in 1954 and 1956 and the ladies stake for five-gaited horses in 1954. Joan Robinson Hill died in 1969 at the age of 38. She had been ill for several days, and her death originally was attributed to pancreatitis. The possibility exists that she was poisoned. Her husband, Dr. John Hill, was tried for "murder by omission" (negligent homicide) in 1971; the case ended in a mistrial. Dr. Hill was shot to death by an intruder at his home in 1972. These events were chronicled in the 1976 bestseller *Blood and Money*, which was written by

Thomas Thompson.

Si Jayne and George W. Jayne both exhibited hunters in competition in 1956. Both brothers married prominent riders: Si Jayne wed Dorothy Macleod, and George's wife Marion often showed her husband's horses. In the 1956 competition, Si Jayne's chestnut gelding, Happy Landing, shown by Dorothy Macleod, won the working hunter $1,000 stake, but relations among members of the Jayne family were far from smooth. The two brothers grew so wary of each other that they traveled with bodyguards. In 1971, Kansas City Star reporter John T. Dauner wrote: "This writer remembers vividly arguing heatedly with Si before he would permit his horse and rider, winners of the American Royal hunter championship to be photographed standing next to George's horse and rider, the reserve champion." Mr. Dauner's recollection appeared at the end of a front-page story: Si Jayne had just been arrested in connection with the death of his brother George who was shot to death at his home in October, 1970. Si Jayne and two accomplices were convicted in 1973 of conspiring to murder his brother. According to some accounts, Si Jayne originally approached one of his former employees at the October 1964 American Royal and asked him to carry out the killing.

On the lighter side, Gene Autry appeared at the 1956 horse show, and Hopalong Cassidy rode in the parade. Entertainment at the 1956 Coronation Ball was provided by Rosemary Clooney, and the 1957 Coronation Ball included a 15-minute performance by Dorothy Lamour. At the 1957 ball, Florence Henderson (later famous as "Carol Brady" of television's "The Brady Bunch") crooned duets with singer Bill Hayes. Jacques d'Ambois and Allegra Kent of the New York City Center Ballet danced to "Shinbone Alley" with the Royalettes, a group of 50 high school girl dancers. Half of the dancers were dressed as black cats; the others were white cats. This preceded the musical Cats by more than twenty years. No doubt many of the BOTARs remember that night fondly — particularly Mariella "Day" Gibson, who later married her escort, Whitney Kerr.

George Shepherd joined the Royal in 1957 and became general manager the following year when C. M. Woodard retired due to ill health. Joanne Faulkner, who was associated with the American Royal from 1963 to 1984, recalled that:

George Shepherd was the General Manager when I started to work at the Royal — a dear, kind, dyed-in-the-wool Texan. It was always rumored, and I have no reason to doubt it, that during some of the especially lean years for the show, he did not take a salary many times so that his employees could take theirs. He probably would deny this — but more than one person has told me this as fact.

George Shepherd, 1964
(Photo courtesy of the Western Historical Manuscript Collection-Kansas City, American Royal Collection).

Olive Lansburgh, who served for a time as horse show secretary and was later president of the BOTAR organization, remembered that Mr. Shepherd always ended the horse show by having the band play "Everything's Coming Up Roses." For several years, Mr. Shepherd and his secretary, Barbara Whorton, were the only full-time employees at the Royal. Maintenance, electrical work and building preparations were performed by Kansas City Stock Yards Company employees. Bill Campbell was in charge of building maintennce. Mr. Campbell was a tireless worker: As Mrs. Lansburgh recalled, "He held the old building together with baling twine and prayers." Among his other activities, he oversaw the kitchen that was set up in the green room at the north end of the arena during the horse show. After Mr. Campbell's death in the mid-1970s, Paul Davis took over responsible for maintenance. C. K. "Smiley" Bell, another Kansas City Stock Yards Company employee, helped supervise the building during the show and was responsible for putting animals in their pens. From the 1940s to the 1960s, shepherds Alex McKenzie of Oklahoma State University, Tommy Dean of Kansas State University and Alvin Dixon of Iowa State University helped out by bringing sheep from their schools, and by acting as judges. Richard Matthews began working year-round at the Royal as a self-described "lackey" in 1958; he served as a handyman and ran errands until his retirement in 1988 at the age of 81. The kitchen in the American Royal offices bears a plaque in his memory. Many volunteers worked on the shows, and the Royal also hired seasonal workers, including a hobo known as "Freight Train," who would help set up the show and then get back on the rails.

C. K. "Smiley" Bell with FFA member Dennis Howard and 1968 champion lamb, junior division (Photo by William L. Glover, courtesy of the Western Historical Manuscript Collection-Kansas City, American Royal Collection).

Cecil L. Eyestone joined the Royal in 1958 as superintendent of the 4-H sheep show. Mr. Eyestone first came to the Royal on Kansas Day in the mid-1930s, when he was a 4-H club member from Leavenworth County. He attended Kansas State University and then became a 4-H agent in Montgomery County, Kansas, bringing livestock judging teams to the Royal in 1947, 1952 and 1957. In 1964, he became the superintendent of the sheep department, replacing Carl Elling, who was associated with the Kansas State College of Agriculture at Manhattan and who retired after 23 years of service. Mr. Eyestone himself retired in 1993.

Among the sheep exhibitors at the Royal in Mr. Eyestone's time were Durham Howard, who raised Southdowns; Dewey Johntz, who showed Cheviots, which are the smallest of the breeds shown at the Royal; Lyle and Lanny Crawford, who showed Corridales for 15 years; Leonard Steward, a Kansan who began breeding Dorsets in the 1940s; Roger Nichols, D. D. S., whose Hampshires were often champions and who also raised Suffolks; Joe Bill Reed and family, who brought their Oxfords to the show; and Silvertop Farm, owners of Shropshires, as well as Suffolk exhibitors Glenn Mortimore and Dwight Stone. Sheep judges over the years have included Harold Barber of Kentucky, Walter F. Renk of Wisconsin and E. L. Shaw of West Virginia. Many sheep show observers fondly recall Amy Wilson from Plano, Texas, who exhibited Hampshires for many years.

Jack Hoffman, who sold his steer for $44,375 in 1946, and his father, Karl, had the champion carlot, 15 Angus steers, at the 1958 American Royal. The carlot was sold to Jud Putsch, one of several restauranteurs, including the owners of the Savoy Grill and the Golden Ox, who vied to purchase champions. For many years, Mr. Putsch annually offered an American Royal dinner at Putsch's 210. Later the tradition was carried on by Plaza III.

In 1959, the *Kansas City Times* noted that: "Wearing Empire gowns of emerald green over taffeta, the BOTARs won applause as they fluttered large fans of white ostrich plumes while varied colored lights played over them." In fact, the dresses that year were quite economical: the gowns were designed by Mrs. Stanley Christopher III and made by a dressmaker for $20 each. Records show that the chiffon cost $0.79 per yard at Emery, Bird, Thayer & Co. and that the taffeta was purchased from Donnelly Garment Co. for $0.55 per yard. Herb Wilson, who headed Emery, Bird, Thayer & Co., donated dresses for some years. Mr. Wilson was president of the American Royal in 1962-1963.

Eddie Williams and Alex McKenzie with prizewinning sheep, 1959 (Photo by Randozzo & Morrison, courtesy of Marie Schubert).

The 1960s

When Jay B. Dillingham went to work for the Kansas City Stock Yards Company in 1937, his father-in-law, Allen Thompson, said, "I just hope you don't get involved in show business." As Mr. Dillingham recalled, "It wasn't a week before I was involved in putting on the Spring Horse Show for the Police Benefit Association." Mr. Dillingham had intended to serve as manager of the stock yards for five years and then return to the practice of law. Instead, he stayed on with the stock yards — and with the American Royal. In 1960-1961, Mr. Dillingham served as president of the Royal.

In the early 1960s, George Powell, who headed Yellow Freight and whose family later endowed Powell Gardens in Kingsville, Missouri, was asked by the American Royal to head a campaign to raise money for a new building, which became the Governors Exposition Building. A group of American Royal supporters had formed the American Royal Association and leased the facilities from the Kansas City Stock Yards Company: This was the American Royal's first step in separating from the stock yards company. As Jay Dillingham recalled, "The first thing Mr. Powell did was to call on me at the stock yards company and frankly say that the American Royal could not raise the money to pay the rent it owed the company." Mr. Dillingham's reaction? "I sure wasn't going to

Miss Silver Mischief, a prizewinner of 1960, with Jay B. Dillingham, American Royal judge Emil Kezac, J. W. Van Natta, and Jack Turner. Mr. Turner owned the Hereford (Photo by William L. Glover, courtesy of the Western Historical Manuscript Collection-Kansas City, American Royal Collection).

61

Loula Long Combs receiving a ribbon in 1958 (Photo by Launspach, courtesy of the Western Historical Manuscript Collection-Kansas City, American Royal Collection.

kick them out." At Mr. Powell's request, the Kansas City Stock Yards Company ended up forgiving the American Royal's debt of approximately $400,000. This was money that would otherwise have been paid out to the company's stockholders as dividends. The Kansas City Stock Yards Company chose to make the continuation of the American Royal its priority.

The era of an icon was ending: Loula Long Combs rode in her last American Royal in 1960. She was driven around the arena in her George IV Phaeton pulled by two Hackney ponies. Even her attire was considered noteworthy: She wore a broad-brimmed hat and a long dress with full sleeves. American Royal veterinarian Bud Hertzog, who grew up on a dairy farm one-half mile from Longview, remembered seeing her show horses: "She always had a little Boston bull terrier that rode in the cart while she was showing. The American Royal horse show ring, which was upstairs in the old building, could seat about 6,500 people. She would absolutely captivate that crowd. There wouldn't be standing room, when Mrs. Combs came into the ring. And you could almost count on her winning." She died in 1971 at the age of 90.

The 1960 beef cattle show was dedicated to Dr. Arthur D. "Dad" Weber, a longtime livestock judge. Dr. Weber was then taking a leave of absence from Kansas State University, where he was Dean of Agriculture and Director of the Kansas Agricultural Experiment Station, in order to head the Ford Foundation's project to increase food production in India. Weber Hall at Kansas State University is named in his honor. It was because of Dr. Weber that Don Good went to Kansas State University as a livestock judging coach in 1947. "Dr. Weber was one of the most respected livestock men in the whole United States," Dr. Good recalled. "He was a progressive man, a good researcher, and an excellent teacher."

Arthur D. Weber demonstrating Hereford characteristics (Photo by Arthur D. Weber, courtesy of the Western Historical Manuscript Collection-Kansas City, American Royal Collection).

Prior to 1960, it was customary for cattle exhibitors to bring a nurse cow, usually a Holstein, which provided milk for calves, to the show. "You had to take care of the nurse cow, plus get the calves to the nurse cow twice a day. On show days, you'd delay that until just before the calf was to go into the ring," Bud Snidow recalled. The idea was to make the calves look more filled-out, but there was concern that the animals were milk-fat. "We wanted to show natural growth and natural finish, so the American Hereford Association led the program to eliminate nurse cows at the shows," Mr. Snidow said.

The Governors Exposition Building replaced the American Royal Annex, after the barn used to house sheep, which was adjacent to the annex, was deemed

That winning feeling . . . (Photo courtesy of the Western Historical Manuscript Collection-Kansas City, American Royal Collection).

unsafe during an inspection. The choice of a name was simplified by the fact that the city had passed a bond issue earmarking approximately $2,000,000 for an "exposition" building. The building, which was almost one-and-a-half times the size of its predecessor, was not quite finished for the fall show in 1962. There was no floor, only some asphalt aisles. The following year, the floor, an office area and restrooms were added.

Joanne Faulkner recalled the old offices: "The Royal's offices were in the Livestock Exchange Building, which was directly in the glide path for the old Municipal Airport. As most of the windows in the building were kept open much of the year, the noise from the cattle pens directly below and the planes overhead often made for some interesting dialogue — especially if you were trying to talk on the phone." She added, "The Livestock Exchange Building was peopled by ranchers out of another era — raw-boned and gnarled livestock men who were always unfailingly courteous."

The fact that the grand champion steer of 1961, an Angus named Maybe II, was purchased by Eddie Williams, who was bidding on behalf of Jud Putsch, is not surprising: Mr. Williams a frequent buyer, both for his company and on behalf of other businesses, and he was married to Mr. Putsch's sister, Eleanor. How-

Maybe II with Col. John Riffle, J. C. Penney, and Judy Vining, 1961 (Photo courtesy of Col. and Mrs. John Riffle).

A champion Shorthorn in the sale ring, 1962. The auctioneer is Ray Simms (Photo by William L. Glover, courtesy of the Western Historical Manuscript Collection-Kansas City, American Royal Collection).

Champion barrow sale, 1963 (Photo by William L. Glover, courtesy of the Western Historical Manuscript Collection, American Royal Collection).

ever, Judy Vining, who raised the steer, attended the Royal with her identical twin sister, Joy. Mr. and Mrs. Williams were the parents of identical twin daughters, Marie and Mary. The Putschs' also had a set of twins: a girl named Ginny and a boy called Bill. All agreed that having three sets of twins directly related to one steer sale was unusual and perhaps unique in the annals of livestock show history.

The lavish beauty of the Coronation and BOTAR balls, set in the midst of a livestock show, has occasionally made for jokes. Andy Griffith, who served as an American Royal Queen judge in 1962, announced, "I'm pleased to be here because everyone looks so expensive." He went on to say that he suspected the fathers of the BOTARs would be "mowing their own lawns for a while after tonight." The Coronation Ball was in fact an unprofitable event, losing up to $30,000 in a single year.

Contract's Commander was champion stallion for five consecutive years, 1962 through 1966. This record is still unbroken. "If you beat Earl Teater, you had to have a good horse," Sug Utz said, "and I beat him with Contract's Commander." Mr. Teater worked for the Francis Dodge Stables in Kentucky, which was one of the largest and finest saddlehorse stables. The biggest thrill Mr. Utz ever had at the Royal, however, was in 1963, when Contract's Commander won over Captain Denmark, which had been world's champion and which was shown by Garland Bradshaw. "My wife, Mary, took my entries to the American Royal on the last day they were due. She called me, and she said, 'We're not going to show Contract's Commander in the stud stake. I'm going to scratch him.' I said, 'Why?' She said, 'Captain Denmark is entered.' I said, 'I don't care if he is. Don't scratch that horse.'" Mr. Utz was determined to put Contract's Commander to the test, and he won.

In 1963, a special mule exhibition was held on Missouri Day. Harry S. Truman rode around the ring on a wagon. During the judging, the band played the "Missouri Waltz." Claude "Brother" Adams of Lamar, Missouri had the two top mules, Jane and Judy. President Truman noted, "Pinned a medal on the prize mule and a half dozen other mules. It was a ribbon in each case that had a rosette at the top that looked like a medal anyway." He jotted this on his copy of the program for the annual Royal 4-H Club which he attended the same evening. Two years later, Charley O., the Kansas City Athletics mascot, made a brief, unscheduled appearance at the Horse Show. Howard Benjamin rode Charley O., who was described by the *Kansas City Times* as "standing in for the ghosts of the hundreds of mules who were once, but are no longer, shown at the Royal."

The Missouri Charolais Breeders Association held their first annual Charolais Congress at the American Royal in 1964. For several years, Jerry Litton served as general manager. Mr. Litton and his father had previously been Hereford breeders. "Jerry Litton brought the Charolais into the Midwest. He was fighting a uphill battle for a while, but gradually the Charolais were accepted," Dr. Don Good recalled. "They played an important role in the development of the beef cattle industry in this country, because they brought more growth and more muscle to the English breeds of cattle." Mr. Litton, who had been the FFA's National Student Secretary in 1956, was a congressman as well as a rancher. He won the Missouri Democratic nomination for the U. S. Senate on August 3, 1976, but was killed in a plane crash that night.

Those who attended the 1965 American Royal Parade would remember it in their bones: There was a sleet storm that morning. "Andy Pat" Paterson retired that fall, after 42 years with the Royal. He was honored with a dinner, at which he became the first recipient of the Walter A. Atzenweiler Memorial Fund Award. Mr. Atzenweiler had been the agricultural commissioner of the Chamber of Commerce in Greater Kansas City prior to his death in 1962. In recent years, his son, Larry Atzenweiler, has sponsored the Walter H. Atzenweiler Scholarship to the Second High Point Individual in the 4-H livestock judging contest; the scholarship for the indi-

Harry S. Truman greeting a mule and its exhibitor, 1963 (Photo courtesy of the Harry S. Truman Library).

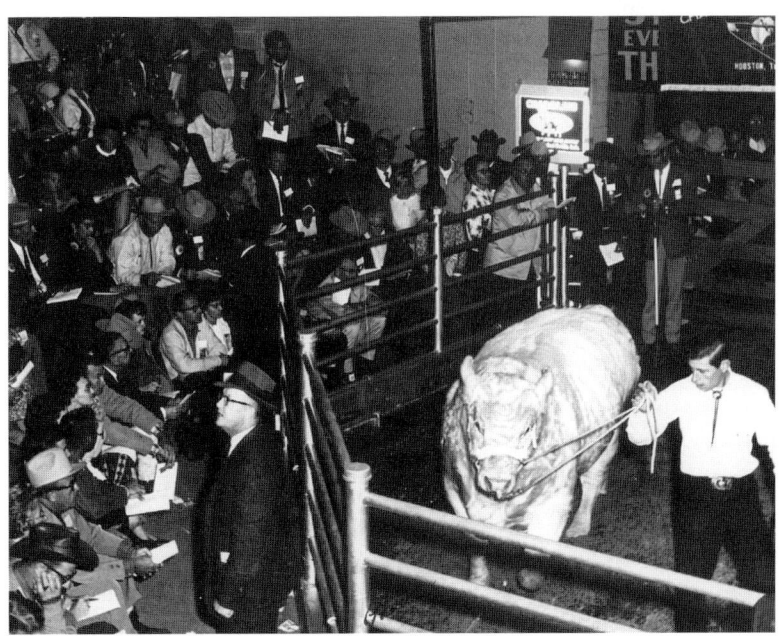

Charolais sale scene (Photo by William L. Glover, courtesy of the Western Historical Manuscript Collection-Kansas City).

Charolais sale scene (Photo by William L. Glover, courtesy of the Western Historical Manuscript Collection-Kansas City).

Petticoat Lane auction, 1965 (Photo by William L. Glover, courtesy of the Western Historical Manuscript Collection-Kansas City, American Royal Collection).

vidual scoring the most points is given by Mary Dickerson Tollefson in memory of her husband, Max Dickerson, who was with the Commercial National Bank. Mr. Paterson was replaced as livestock show manager by Dr. Raymond L. Burns.

Also in 1965, the champion steer was sold at a livestock auction held on Petticoat Lane instead of in the Royal arena. New that year was the National Championship American Paint Horse Show. One regional newspaper misunderstood a lecturer's comment, and announced that the Royal would feature a "pink horse show."

The Royal Queen was Deborah Fowler, a junior at the University of Kansas. The *Kansas City Star* recorded that "The 1965 queen entered the Arena in a Fayette Boot Victoria coach drawn by four silver-dappled shetland ponies." The coach was 114 years old; it was owned by Elmer C. Adams of Blue Springs, Missouri and driven by E. C. Adams, Jr. Of the Queen's attire, the *Star* wrote: "Miss Fowler's gown was of white English net over white satin. Beads and pearls were sewn in [a] floral pattern." The BOTARs wore white satin gowns with pearl beading. Wayne King, the orchestra leader, enthused, "This is the most gala event of the autumn in the entire United States; it should be televised all over the country."

The rodeo was revived as a spring activity in the American Royal Building from 1965 through 1974. Hank Walton surprised American Royal officials when he won the four-county pony express race at the American Royal Rodeo in the summer of 1967. The flagman was at dinner when Mr. Walton rode in at 7:15 p.m., 15 to 45 minutes ahead of schedule. Bob Jones, public relations man for the Royal, was photographed waving Mr. Walton in with a handkerchief. The program for the evening included calf-roping, saddle bronc riding, steer wrestling and wild cow milking. Michael Landon, who played "Little Joe" on the television show *Bonanza,* sang and rode around the arena.

Melissa "Missy" Newby (now Shores) was crowned American Royal Rodeo Queen that night. "It was so exciting, everyone was screaming and yelling," she recalls. She previously had shown 4-H club calves and quarter horses, once winning the reserve championship. It was at the Royal that she first met horse exhibitor Roger Shores, whom she subsequently wed.

The Knights of the American Royal was started in the mid to late 1960s. It was based on the Knights of Ak-Sar-Ben at the Omaha racetrack. Members paid dues, and raised money for entertainment. "The whole idea was to get some support for the Royal, have a good time and have a big-name entertainer come to town," Bob Hovey remembered. Tennessee Ernie Ford was one of the first performers hired by the group. Unfortunately, the spotlights momentarily blinded him, and he fell off the stage. Being a veteran trouper, however, he insisted on finishing the show. The Knights of the American Royal soonafter disbanded.

Arthur Godfrey, 1940s radio personality and television pioneer, and his gelding, Goldie, entertained at the horse shows in 1966 and 1967. Mr. Godfrey was an accomplished horseman in dressage. Although he was known to be a difficult man to work for, the audience at the Royal saw a different side of him: He was happy to stay after the show and talk with members of the crowd. He was an enthusiastic supporter of the horse show, recommending a visit to the Royal to radio and television audiences.

The Hereford Association held a conference at the American Royal in 1967 and decided to work toward developing "longer, leaner, larger" animals. This would cause tremendous change in the cattle industry, as would the popularity of imported breeds, beginning with Charolais in the 1960s and continuing with Chianinas and other breeds in the 1980s. Cattle judges of the past 35 years have included Herman R. Purdy, Dr. Don Good, Dr. Dave Nichols, Dr. Miles McKee, John McKnight, Harlan Ritchie, Dr. Gary Minish, Dr. Robert Totusek and John C. "Jack" Ragsdale. Judges from other countries who have participated include Walter and James Biggar of Scotland, Charles Yule of Canada, Carlos and Charlie Duggan of Argentina and German Morixe of Uruguay.

American Royal Queen Deborah Fowler, 1965 (Photo courtesy of the Western Historical Manuscript Collection-Kansas City, American Royal Collection).

George Rush, who had been a young Kansas City Stock Yards Company employee at the time of the 1951 flood, became the Royal's accountant in 1967. He held that job until he retired in 1996. His assistant for 25 of those years was Evelyn Crews.

King Lee, a 17-year-old gelding belonging to Janet Green was retired at the American Royal in 1967. The five-gaited horse made his final rounds of the show ring with his owner, and then posed for photographs with Miss Green, Senator Edward V. Long of Missouri and Art Simmons, who had been King Lee's trainer since 1958. During his career, King Lee won eight five-gaited classes for women and amateur riders at the Royal.

Harry Darby was elected chairman emeritus of the American Royal in February, 1967. His work, however, was far from finished. The following year, Senator Darby, Herbert H. "Herb" Wilson, chairman of the Board of Governors, Roderick "Rod" Turnbull, American Royal president and editor of the *Weekly Star Farmer*, and E. K. "Joe" Hartenbower, general manager of KCMO, who had served as American Royal president in 1964-1965, sent out invitations to a "Mortgage Burning Dinner" at the Muehlebach Hotel at 7 pm on May 11, 1968, with entertainment to follow at 8:30 pm at the Municipal Auditorium Arena. The purpose of the dinner was to pay for expenses incurred in the building of the Governors' Exposition Building. The price was high: $100 per plate. But it was worth it: Bob Hope performed. Mr. Hope was, as one might suspect, a friend of Senator Darby.

The Children's Art Show began in 1968. Olive Lansburgh (then Olive Wright) co-founded the art show with Rosemary Beymer, who was director of art education for the Kansas City, Missouri public schools. "One thing that has changed is that they now have winners," Mrs. Lansburgh said. In the early years, twenty blue ribbons and twenty red ribbons were awarded. Forty children could then ride on the float in the parade. The awards were given by the children's art teachers because Ms. Beymer believed that she and Mrs. Lansburgh would put too much emphasis on factual details rather than artistic expression. One picture Mrs. Lanburgh never forgot showed a horse wearing shoes: bright red high heels, to be exact. Since the mid-1990s, the Royal has also sponsored a youth photography contest for grades 10 through 12.

In 1969, the FFA voted to allow girls as members. Julie Smiley of Mount Vernon, Washington became the first female national officer in 1976, when she was elected National Vice President of the Western Region. The Star Farmer of America award had been given since 1929; in 1969, The Future Farmers of America Foundation created the Star Agribusinessman award to honor outstanding candidates in nonproduction careers.

Fess Parker, who had played Daniel Boone and Davy Crockett on television was on hand for the 1969 American Royal. Sue Noll recalled, "We were still in our old offices in the Royal building. I stepped out into the hall. There was Fess Parker standing in the middle of a circle of blind children, going from child to child, shaking hands and speaking to each of them."

American Royal Children's Art Show (Photo by Melissa Shores, courtesy of the American Royal Association).

The 1970s

The 1970 American Royal Register of Merit Hereford Show honored Donald Ornduff, who retired from the American Hereford Association that year. He had joined the staff in 1930, and became editor of the Hereford Journal in 1944.

Sheep judges since the early 1970s have included Dr. Carl Menzies, Dr. Leroy Boyd, Tom Durham, Merrill Neary, Dr. Jack K. Judy, Ron Guenther, Hal Yeager, Dr. Hilton Briggs, Dr. Merle Light, Darrell Anderson, Duron Howard, Mike Caskey, Rod Crome, Jamie Farao, Ron Young, Bill Crutcher, Kenneth Urban, Dale Smith, Jat Mittag and Tom Clayman.

Roy Rogers and Dale Evans appeared at the Royal in 1971, as did the Sons of the Pioneers; The Royal Canadian Mounties performed in 1972. It was not the first trip to Kansas City for the Mounties, who have pleased several generations of audiences at the Royal, including crowds at the 1958 and 1972 shows. The Mounties are scheduled to return for the centennial.

J. Ralph "Pistol" Peak became horse show manager in 1972 and retired after the 1978 show. He was the third generation of the Peak family to be in the horse business; his sons George and Sam followed the family tradition. Prior to joining the Royal, he had served as manager of the Illinois State Fair, superintendent of grain inspection for the State of Illinois and manager of the Fernwood Stables of Barrington, which was owned by the Beuhler family.

There were some bleak days for the American Royal in the early 1970s. The American Royal

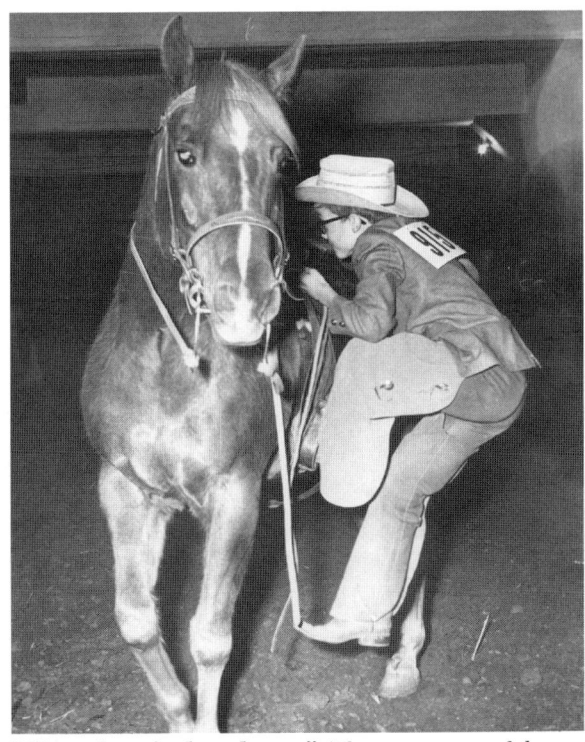

"Okay, here's the plan..." (Photo courtesy of the Western Historical Manuscript Collection-Kansas City, American Royal Collection).

The 1970s

Building had deteriorated over the years. Bud Snidow recalled that, back in the 1950s, the floors developed cracks when trucks loaded with sand were brought to the second floor. The sand was meant to serve as bedding for cattle. The cracks were temporarily shored up with timbers.

The Bit and Spur Club, a cave-like bar by the horse show arena, was a popular hang-out for riders, exhibitors and spectators. It was located under the driveway off the 23rd Street Viaduct, and it was frequently both damp and dark. "It got so bad that when it rained, the water would run in. We hung parachutes to catch the water. You had to be careful, because if you hit one, the water spilled," Bob Hovey recalled, laughing.

The acoustics in the building were not good: Two decades later, people still remember how the roof reverberated when the Wentworth Band from Lexington, Missouri played. Everyone still enjoyed the show, however, just as they did when students from the Culver Military Academy in Culver, Indiana performed with their Black Horse Troop.

Worse, the American Royal was not paying the rent. The property was owned by the Kansas City Stock Yards Company and leased to the American Royal. Over the years, Jay Dillingham and the Kansas City Stock Yards Company stockholders had allowed the Royal to continue to occupy the building without paying, but the situation could not continue indefinitely.

As Bill Harsh recalled, "John C. Gage was president of the American Royal in 1972. We were

Royal Canadian Mounted Police Equitation Team (Photo courtesy of the Royal Canadian Mounted Police).

just about to go under. We had no money. We owed the stock yards company for rent . . . I think about $380,000. We owed everybody else. John Gage wrote a letter that went to the Board of Directors and the Board of Governors, and maybe to others, and told them of the situation we were in. He said 'I may be the president who is going to preside over the death of the American Royal.' That upset a few people no end. It was the right letter."

John C. Gage recalled, "Some people were very angry with me. But fortunately, some of the more level-headed ones — and Bill Harsh was one of them, God rest his soul — jumped in with Bill Theis to help solve the problems."

To add to the Royal's problems, there was a feeling among some Kansas Citians that the presence of a livestock show did nothing to enhance the city's image. "Before my father died, I went to the Civic Council [a consortium of business leaders]. I said that I would give the leadership to raise the money to buy the complex from the stock yards company if they would help me," R. Crosby Kemper, Jr. remembered. "There had been a study made in New York that we should get rid of our cowtown image; we should be a city of science and education and fountains. I thought we could have both: science, education and fountains, and our roots and our history." The Civic Council was not interested in Mr. Kemper's proposal. "I asked the membership-at-large who would help me? Bill Harsh was the only guy who raised his hand."

William P. "Bill" Harsh served as president of the American Royal in 1975-1976 and as chairman of the American Royal Horse Show from 1975 to 1991. He was instrumental in developing the horse show into a nationally recognized event. Mr. Harsh grew up around horses: His father, a saddle horse judge, gave him his first pony when he was three years old. Mr. Harsh was an executive with Hallmark Cards, and also was known as "Mr. American Royal." He was named "Mr. Kansas City" by the Chamber of Commerce. He is remembered as a gentleman by all who knew him. Shirley Parkinson recalled, "He came through the hallways at night, welcoming everybody, making sure everything was all right. Most of the officials park in lot B. He never did. He always parked in lot A, so that he

Sheep judging scene, 1971 (Photo by William L. Glover, courtesy of the Western Historical Manuscript Collection-Kansas City, American Royal Collection).

Champion hog with exhibitor, 1971 American Royal Queen Debbie Lee Carey, and onlooker (Photo by William L. Glover, courtesy of the Western Historical Manuscript Collection-Kansas City, American Royal Collection).

William P. Harsh receiving the UPHA Horseman of the Year award from UPHA President Charles Crabtree (Photo courtesy of the American Royal Association).

R. Crosby Kemper, Sr. (Photo by Strauss-Peyton, courtesy of the American Royal Association).

could walk back through the barns and the Governor's Building to his car."

Mr. Harsh encouraged the additions of the UPHA (United Professional Horsemen's Association) Classic Grand Championships and the AHSA (American Horse Shows Association) Medal Finals. He helped bring the Concert for Champions to Kansas City. He also was an enthusiastic supporter of the construction of the Governor's Building and the new American Royal Building. His son, Hall Harsh, has served on several American Royal committees.

Mr. Kemper's confrontation with the Civic Council had far-reaching consequences: He resigned from the council, and the Midwest Research Institute did a study on the American Royal. "The study showed that the Royal was our most important asset," Mr. Kemper recalled. At that point, Charles Kimball, who headed the Midwest Research Institute, the Hall Family Foundation and several other groups became involved in efforts to save the Royal.

On October 24, 1972, R. Crosby Kemper, Sr. died. George Shepherd recalled, "Mr. Kemper was one of the finest people. He was our treasurer for many years and was offered the presidency. He said, 'No, I'll stay treasurer.' He was always willing to help." In 1923, Mr. Kemper told a reporter from the *Kansas City Star* his vision for the future:

> I want to see Kansas City become a city of a million people and be recognized as the outstanding city in a commercial way of the entire West.
>
> I want to see more manufacturing plants locate here, and to have the trademark, "Made in Kansas City," known all over the United States.

Shortly after Mr. Kemper's death, R. Crosby Kemper, Jr. announced that the family trusts would give $1.5 million to the American Royal. This gift would begin the events that would culminate in the creation of Kemper Arena.

A bond issue passed by voters in 1954 specifically earmarked $5.6 million in general obligation bonds for a sports arena and exposition center that would house the American Royal. In order for the funds to be released for the project, the City Council had to approve the proposal.

Originally, the proceeds of the bond issue were to have been used to develop a site on the Missouri River for the Royal, but those plans had been scrapped. Conflicts remained: Some people wanted to combine the Royal with the Truman Sports Complex, others proposed moving it to 111th and Switzer in Kansas. Neither proved feasible. A group of American Royal officials and livestock breeders inspected Union Station and decided that it would be prohibitively expensive to create a

Kemper Arena (Photo courtesy of the American Royal Association).

combination American Royal and sports arena on the site. As Kansas City Star sportswriter Joe McGuff wrote at the time: "What Kansas City needs least is another arena proposal. In the last two years 11 arenas have been proposed on eight sites in Greater Kansas City."

Mr. Kemper felt that the Royal should not be moved. "The people voted the bonds for an American Royal arena," he said. "They voted for the American Royal. When they were asked to vote for a sports arena, they voted it down." Two possible tenants for the new arena were the Kansas City Kings basketball franchise and Edwin Thompson, who held the National Hockey League franchise for Kansas City and who had until Thursday, January 25, 1973 to file a proposal for a site and financial backing with the N.H.L.

The Royal would be saved — but it was already obvious that more funds would be needed. "Bill Theis and I went to talk to Crosby on Christmas morning," Mr. Gage recalled. "The city thought it could provide bond financing of somewhere around $10 million, but we needed to come up with $10 million to go with that." Mr. Kemper was quick to respond: The Kemper donation was increased to $3.2 million dollars.

As soon as the additional funding from the Kempers was announced, the American Royal Board of Governors voted to exercise its option to buy its own site. The board offered $2.5 million to the Kansas City Stock Yards Company for 35 acres of land, which included the American Royal building, the exposition hall and a parking area as well as property occupied by the Sutherland Lumber Company.

Additional financing for the Royal was being arranged. Bill Harsh, assisted by Jerry Scott, headed the drive to raise funds in addition to those provided by the Kemper family. The goal was $1.5

million. In fact, the men raised $2.7 million. Richard Stern and John F. Fogarty, Jr. of Stern Brothers created a plan whereby Stern Brothers would underwrite $7.5 million in revenue bonds for the project.

Certainly, everyone involved wanted the best for the Royal's future. In spite of this, the deal nearly fell apart. On Friday, January 19, 1973, after the Kansas City Stock Yards Company and the American Royal failed to come to an agreement on the price of 21 additional acres, the board of directors of the American Royal Association asked the City Council to condemn 35 acres of land belonging to the stock yards company. For years, the Royal had operated under the auspices of the Kansas City Stock Yards Company; now the two were on opposing sides. That same day, the City Council voted — and came within one vote of defeating the proposed arena. Mr. Fogarty declared that Stern Brothers would not purchase the $7.5 million revenue bonds.

On Monday, January 22, 1973, the situation was resolved. The City Council met again. R. Crosby Kemper, Jr. announced that the American Royal Association and the Kansas City Stock Yards Company had reached an agreement: The Royal would purchase 53-1/2 acres of land for an undisclosed sum. As a result, Stern Brothers had decided to buy the revenue bonds. *The Kansas City Times* quoted Mr. Kemper as saying, "This is contingent . . . on you gentlemen voting these bonds this afternoon. If you don't, the whole deal is off. We withdraw our gift, Stern Brothers withdraw their agreement and we forget the whole thing." The City Council held an informal vote in favor of issuing the $5.6 million in general obligation bonds.

Kemper Arena, which was designed by the Chicago firm of C. J. Murphy, was dedicated on October 18, 1974. Earl Butz, Secretary of Agriculture under President Ford, attended the opening ceremonies. "It was a thrill to have the Secretary and his wife take time to walk around the livestock exhibits. Before he departed, he walked into a few pig pens along with young members of the FFA," Bill Theis, American Royal president 1973-1974, remembered. "The American Royal . . . this prestigious show is a major connecting link between the city's trade and transportation facilities and the agricultural empire that surrounds them," Secretary Butz said. "The American Royal is one of the places where we compete for breed excellence and for improvement in the breeds . . . It is a place for introduction of new methods of breeding and management."

The following month, President Gerald R. Ford himself came to town. He gave a speech on the economy and his WIN (Whip Inflation Now) campaign to the FFA convention. President Ford returned to Kansas City in July, 1976, when the Republican National Convention was held at Kemper Arena. R. Crosby Kemper, Jr., opened the convention and presented a Missouri walnut gavel to chairwoman Mary Louise Smith. The arena is a multi-purpose building, and has frequently been the site of concerts and sports activities.

In 1975, the first year that the American Royal was held in Kemper Arena, the rodeo became part of the American Royal's livestock and horse show events in the fall. The American Royal Rodeo is officially sanctioned by the Professional Rodeo Cowboys Association and by the Women's Professional Rodeo Association. Entertainers perform between competitions. The musical acts are booked months in advance, and the Royal has made some remarkably prophetic choices. One of the first groups to play was a band called Alabama, which was virtually unknown when it was signed, but was a very hot ticket by the concert date. When Crystal Gayle appeared in 1977, her song "Don't It Make My Brown Eyes Blue" was at the top of the charts.

Will Shriver was the five-gaited champion of the American Royal. He was the first horse foaled in Missouri to win the event. The three-time world's champion stallion was owned by Mr. and Mrs. Weldon's Callaway Hills Stable, trained by Redd Crabtree. In his career, Will Shriver won the five-gaited stallion and championship division stake at every major show in the United States. Will Shriver's family has a distinguished history at the Royal. His dam, Kate Shriver, trained by Garland Bradshaw, won the fine harness class. His sire, Johnny Gillen, was the reserve champion in the

Rodeo scenes: top, Erik Totten, bottom left, Martha Josey, bottom right, Casey Murphy
Next page: top, Neil Hummel in mid-air as Leon Coffee distracts Steiner's "RL," bottom, Junior
Lewis (Photos © Bern Gregory Photographic Collection, accession no. 99.025, National Cowboy Hall of Fame, Oklahoma City, OK).

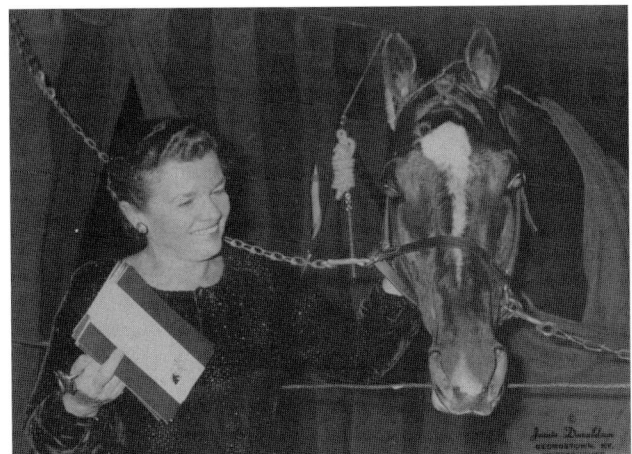

Betty Weldon and Will Shriver (Photo © by Jamie Donaldson, used by permission, courtesy of Betty Weldon).

Wish Me Will and rider Redd Crabtree receiving a ribbon in 1996 (Photo © by Howard Schatzberg, used by permission).

Missouri-Kansas Stake when he was shown by Dale Pugh. Johnny Gillen was sired by Wing Commander out of Fourth Estate, who was the first horse that Betty Goshorn (later Betty Weldon) ever owned. (The name "Fourth Estate," which is used as a synonym for the press, seems quite apt given that the Goshorn family owns the News Tribune Company. Mrs. Weldon named Will Shriver in honor of her husband, William Weldon.) Will Shriver retired on November 20, 1976. "It never occurred to me to retire him any place except the American Royal," Betty Weldon said. The 1991 Concert for Champions was dedicated to him. At the 1996 American Royal, his son, Callaway's Wish Me Will, shown by Mr. Crabtree, added his name to the list of five-gaited champions. In 1999, as part of the American Royal centennial celebration, the Breyer Company is marketing a Will Shriver horse. The replica, like the original, will have a red braid in his mane. One attribute of the stallion the Breyer company was unable to duplicate: Mrs. Weldon (also known as "Will's Mom") has said that Will Shriver loved peppermints, and always could tell if she was carrying some in her purse.

In 1977, George Shepherd retired and was replaced as general manager by Laurence Pressly. When J. Ralph Peak retired as horse show manager in 1978, it was announced that Alan F. Balch would succeed him. Instead, Mary Lou Funderburgh became the horse show manager in 1979. Mrs. Funderburgh had served as ring secretary in 1969, and had also exhibited saddlebreds.

Bob Hovey, who was president of the American Royal in 1979-80, will always remember the first week of June, 1979: The roof of Kemper Arena collapsed, and the Kansas City Club, of which he was president, had a fire. Fortunately, the roof collapse occurred on a night when the building was unoccupied: Half of the roof panels dropped, falling into the seats. It quickly became obvious that the repairs would not be finished in time for the Royal. What to do? "We got to have the show in the old arena, the arena where I had grown up with the American Royal, with the walk-around and the boxes. My wife Carol and I sat in the President's Box, which was in the center on one side," Mr. Hovey recalled. "We knew it was the last time."

Increasingly, the Royal branched out into new areas. "One day, I had been to a sale of western art, and I said to Lawrence Pressly, 'Why don't you start a western art show?' And they did," Bud Snidow recalled. Ray R. Evans, who loved western art, founded the show and served as its chairman for several years. Artists featured included Joe Abbrescia, James Boren, Robert "Bob" Dorman, Sally Jackson, Judy Mackay, Rusty Phelps, Tom Ryan and Grant Speed. Tom Beard, Mary Campbell, Cindy Fauntz, Jim Hamil, Bernard Martin, Jack O'Hara, Rich Rudish and Bud Snidow were among the local

BOTAR officers, 1976, left to right: Jan Coulson, activities chairman, Georganne Hall, ball chairman, and Sharon Franke, president (Photo by Ken Taylor, courtesy of the BOTAR Organization).

artists who participated. George Morse, an enthusiastic supporter of the Royal and of the art show, served as auctioneer in the early years of the show. In 1979, an American Royal art show was held in an upstairs room in the arena. The art show later went to the Alameda Plaza Hotel, where it was held for several years. It next moved to a downtown bank building. During this period, Alexander C. "Sandy" Kemper served as chairman and worked to involve young people. The most recent art show was held in the American Royal complex.

The 1980s

1980 saw the beginning of the American Royal Barbecue, which has since flourished. This should not be surprising, given that Kansas City is renowned for its barbecue. George Zahn organized the first barbecue at the request of Bob Hovey and Laurence Pressly. He was assisted by Bud Mischler, who owned the Bum Steer Bar-B-Que in Lawrence, Kansas, and Tom Stapp, who had previously volunteered his services at the barbecue contest hosted by the Houston Rodeo and Livestock show, and by an eager group of committee members, most of whom had never attended a barbecue contest. The barbecue was held on the showgrounds and attracted approximately 15 to 30 chefs. The rules stated that "Contestants must cook a minimum of ten (10) pounds of barbecue which may consist of beef, pork or lamb. No single steaks will be considered by the judges. Only one cut of meat to be judged per contestant or per team." Some participants defied the orders and made sausage. The judging was held on Saturday, November 8th at 4 p.m., and the bluegrass band Riverrock played. Judges, who included Mike Murphy of KCMO-Radio and sportscaster Jack Harry and anchor Wendell Anschutz from KCMO-TV, were asked consider to appearance, taste, aroma and texture, and to rate the fare on a scale of 10 ("Super Excellent") to one ("Not For Me!"). Psychiatrist Rich Davis was thinking about giving up his medical practice in favor of marketing his own "K. C. Masterpiece" barbecue sauce. The judges thought that was a fine idea, and awarded him first prize.

The barbecue moved to an area directly behind the Golden Ox Restaurant the next year, and returned to that site after a brief foray to Crown Center. Calvin Trillin wrote of the equipment on show at the 1982 barbecue:

> In the Crown Center plaza, people were setting in place huge rolled-steel drums with fireboxes welded to the sides. The barbecue rigs were every conceivable shape; some of them were so large that they had wheels, so they could be hauled behind pickup trucks. The team of a local artist named John Puscheck stood out in the equipment category: it somehow managed to barbecue in white porcelain stoves that looked as though they had been rescued from the demolition of a nineteen-forties apartment building. The meat was put in the oven, the fire was built in the pots-and-pans drawer underneath it, and smoke eerily out of the holes where the burners had been.

By the mid-1980s, the barbecue was in full sizzle. Robert C. Trussell recorded the preparations for the 1985 event in the *Kansas City Star*:

> Tonight more than 70 barbecue cooks, amateurs and professionals, will descend on the parking lot behind the Golden Ox restaurant at 1600 Genessee Street, with a convey of smokers and grills and hundreds of pounds of animal flesh... Each one [of the 72 barbecue judges] will have to taste as many as 15 samples, using only a little white bread and beer to "cleanse the palate" between tastes.
>
> "You have to be a brain surgeon, a jet pilot, an all-around nice fellow and a pinch hitter," said judge Bob Lockett.

In 1998, more than 340 teams competed in "The World's Largest Barbecue."

Earl Smith, then chairman of the ticket committee, and Tom Allen created the American Royal Ambassadors in 1980. Originally, it was composed of a small group of American Royal supporters who made calls and sold season tickets and renewals in the early spring. In 1998, the Ambassadors sold more than $475,000 in season tickets, governorships and sponsorships. Ed Setzler, Tom Scott and Mary Ellen Purucker were the year's top salespeople.

The success of the steer auction breakfast inspired others. Cissy and Don Wheelock and Vicky and Gary Smith and Missy and Roger Shores began giving a wine and cheese party before the lamb auction.

Linda and H. Michael Coburn at an American Royal Barbecue (Photo by Jere Kimmel, courtesy of George Zahn).

The party was held annually from 1980 until 1997. In 1981, Linda and Mike Coburn, Chris and John Wheat and Sandy and Gary Calvin followed suit with a party prior to the hog auction. Over the years, many people were active in organizing and attending the gatherings.

In 1983, Bud Snidow retired as assistant secretary and director of breed activities for the American Hereford Association. In recognition of his 32 years of service to the association, the 1983 American Royal Register of Merit Hereford Show was named the "Bud Snidow Royal." H. H. "Hop" Dickenson, executive vice president of the American Hereford Association, told the *Hereford Journal*: "Probably no one in the history of the breed has had closer contact with and been friends with more Hereford breeders than has Bud. His calm good nature, willingness to listen to and work to help solve the problems of others, combined with his friendly attitude and overall knowledge of the industry, has endeared him to all cattlemen."

The Shelby County Chips cooking station at an American Royal Barbecue (Photo by Jere Kimmel, courtesy of George Zahn).

Lining up at the horse show, 1981 (Photo courtesy of the Western Historical Manuscripts Collection-Kansas City, American Royal Collection).

Don Harris and Imperator completing a victory lap after winning the 1982 Five-Gaited Grand Championship (Photo © by Tod Macklin, used by permission).

1983 Parade of Champions, left to right: Harlem Globetrotter and Larry Hodge, Night of Roses and John Biggins and Simply Blues and Debbie Foley (Photo © by Tod Macklin, used by permission).

James D. "Jim" Taylor (Photo courtesy of the American Royal Association).

In 1983, the American Royal Swine Show-man of the Year Award was established. The 1989 winner was Ray Masters, who had been an exhibitor at the Royal since the 1930s, but had never had a grand champion.

Mo, a barrow belonging to Jennifer Dunn, 10 year-old daughter of Mr. and Mrs. Leon Dunn of St. John, Kansas, won the grand championship of the market barrow class in the open division in 1984 – "something no other Kansas hog has done for many years," according to the *Garden City [Kansas] Telegram.* According to Cecil Eyestone, it was also very unusual for such a young exhibitor to take first prize.

The American Royal was having a rocky time: The show had lost money six out of seven years, losing $130,000 in 1984. Attendance had dropped as well. David Rismiller, president of the American Royal, was among those who felt that management changes were in order. George Shepherd came out of retirement to manage the 1985 show. James D. "Jim" Taylor, who had served as general manager and secretary of the Iowa State Fair, was hired as executive vice president and general manager in October, 1985.

A horse remembered by many in the audience at the Royal was Joe II, better known as Kickin' Joe. People loved to see him swish his tail and kick one leg out as he took the jumps. Kickin' Joe, ridden by his owner, Bob Kraut, won the Roy A. Edwards, Jr. Memorial Grand Prix in 1988. After Kickin' Joe died in 1992, Mr. Kraut donated the Joe II Memorial Trophy. The first

Jennifer Dunn with champion barrow Mo, 1984 (Photo by Jim Jones, courtesy of the American Royal Association).

American Royal president Dwight Sutherland receiving a check from BOTAR president Janelle Coulson, 1984. American Royal general manager Laurence Pressly stands between Mr. Sutherland and Mrs. Coulson. Robert D. Hovey is at right (Photo by Strauss-Peyton, courtesy of the BOTAR Organization).

recipient of the trophy was Rusty Holzer, who rode Picasso to victory in the 1992 Roy A. Edwards, Jr. Memorial Grand Prix. The Kraut family has had several Grand Prix wins: Laura Kent Kraut placed first aboard Simba Run in 1991 and aboard Sunday II in 1996.

The first "Concert for Champions" was held in 1988. *The Kansas City Star* quoted Susan Franano, manager of the Kansas City Symphony: "I think it's safe to say this a real change of pace for both the Kansas City Symphony and the American Royal." William McGlaughlin conducted, and the evening began with "Overture to Candide" by Leonard Bernstein. An American saddle-bred stallion was paired with

Bob Kraut and Kickin' Joe (Photo © by Jack Schatzberg, used by permission).

the "Toreador Song" from Georges Bizet's *Carmen*. Richard Wagner's prelude to *Die Meistersinger* was the background for a performance by a Morgan stallion. Music included the "Orange Blossom Special" as well as compositions by J. S. Bach, John Philip Sousa and Aaron Copland. Joey Straube and Pam McKee, both former BOTAR presidents, and William Harsh were the founders of the black-tie event, which is modeled on a "Concert for Champions" held at a horse show on the East Coast. Joseph G. Pfeffer wrote this description of the 1990 Concert for Champions, the first one performed by the Kansas City Youth Symphony under the direction of Dr. Glenn Block:

> Opening night spectators were treated to Don Harris and Shoobop Shoobop racking and trotting to music from *Les Miserables*. Taja Bleu Setzer, possibly the busiest exhibitor this week, demonstrated saddle seat equitation on World's Champion Champagne Steffi to the stirring Slavic melodies of Aram Khachaturian. Leslie Erickson and CH Harlem's Diamond Jim danced three gaited style to the theme from *Lawrence of Arabia*. Big Brown Bess, right back in the ring on Monday Night, did the fine harness park trot to "Somewhere Out There" from *An American Tail*. Sultan's Main Event and Alice Sias demonstrated pleasure horse form to the pop standard "More."

Dr. Block and the Kansas City Youth Symphony have since performed annually. The "Concert for Champions" is now an American Royal tradition.

American Royal Student Ambassadors Lisa Killpack and Stephen Walter Cline (Photo courtesy of the American Royal Association).

The 1980s

In 1988, South Dakotan Nicole Sittner would be the last American Royal Queen. Beginning in 1989 one young man and one young woman would be chosen for the American Royal Student Ambassador Program. The candidates were originally from the FFA. As of 1998, the competition is now open to any college student studying agriculture. The ambassadors are selected for their academic achievement and commitment to agriculture. The winners appear at the American Royal. As part of their prize, they receive paid internships and money for college tuition.

Bill and Pam McKee, Karen and Jim Taylor, Jo Straube and Lawson Jones at the gala preceding the 1998 concert for Champions (Photo by Walter J. Fink).

A polo exhibition at the Concert for Champions (Photo © by Howard Schatzberg, used by permission).

The Concert for Champions (Photo © by Howard Schatzberg, used by permission).

The 1990s

In 1990, Marion Vande Wall became the horse show manager. The following April, the United Professional Horsemen's Association named the American Royal Horse Show as the National Honor Show for 1990. Mr. Vande Wall was selected as Horse Show Manager of the Year. He received the first annual Herman R. Miles Memorial Award. Mr. Miles had been a steward at the American Royal. The award is given by the United Professional Horseman's Association.

The last cattle sale at the stock yards was held on October 31, 1991. Many factors contributed to the stock yards' decline. As the years went by, the older packing plants had become outmoded, and closed. Refrigerated trucks had replaced the railroads, and made shipping livestock to a central point unnecessary.

President George Bush addressed the FFA on November 13, 1991. "I cannot give up on the quest for peace on earth," he said. "I owe it to your generation and the next generation to continue to lead, to use America's moral leadership to that end." The president received the third Outstanding American Award ever bestowed by the FFA.

As the stock yards were winding down, the American Royal, which was by then being operated by the American Royal Association, was preparing for a rebirth. In November, 1987, animals at the Royal began to suffer from pneumonia. It was nothing new for exhibitors and spectators to complain of colds: The ramp in the old American Royal Building had been known as "Pneumonia Alley." Kansas Citians made jokes about "American Royal Pneumonia," while out-of-towners called their malady the "Kansas City Crud." But this was more serious. A thorough examination revealed that the old building was falling apart: bad ventilation, a leaky roof and falling plaster. Earlier, the flooring had been replaced, a project that had been overseen by Dwight Sutherland and city officials. At this point, however, more extensive work was needed.

"The executive committee at that time started working with the Jackson County executive Bill Waris," Malcolm M. "Mick" Aslin recalled. Ray Evans, president of the Royal, and other members of the committee supported a Jackson County capital improvement plan, which would have provided financing for the Royal, but voters rejected the $495 million package.

Draft horses at the American Royal in the 1990s (Photo © by Robert L. Pierce).

Charles W. "Bud" Keller served as American Royal president from 1989-1990. Mr. Keller, an engineer with Black & Veatch, also served as chairman of the American Royal facilities committee. Working with the committee on the design for the new American Royal building were Billy Dean Wunsch of Black & Veatch Architects and Tom Bean, city architect for Kansas City, Missouri.

The trickiest aspect, however, was not design but financing. As Mr. Keller recalled, "A city councilman, Emanuel Cleaver, began to take an interest in the Royal's problems." Councilman Cleaver, later the first African-American to serve as mayor of Kansas City, Missouri proposed using an existing sales tax authorization to service zero coupon bonds in the amount of $100 million. The Royal was to receive $25 million, (later reduced to $20 million), with the remainder going to projects such as the Kansas City Jazz Museum and flood control for Brush Creek. The Cleaver plan passed the City Council.

"The original plans called for a fairly basic barn building," Mr. Aslin remembered. He met with Mayor Richard Berkley, city manager Dave Olson and Mr. Bean, the city architect. "The city manager and mayor were very involved in those early discussions. There was a fair amount of give and take," Mr. Aslin said. City officials expressed a desire for a multipurpose facility, which would be significantly more expensive than the building the Royal had been considering. As Mr. Aslin recalled, "The price went up from $18 or $20 million to $26 million. The Royal was committed to raise

$4 million, which increased to $6 million and subsequently to $8 million, when the museum was added."

Mick Aslin headed the capital campaign, and R. Crosby Kemper, Jr. worked closely with him. James and Jonathan Kemper of Commerce Bank stepped forward with $1.25 million for advance planning. These funds came from the William T. Kemper Foundation. William T. Kemper had been an avid fan of the Royal; his scrapbooks, which include American Royal clippings and memorabilia, are now in the collection of the Jackson County Historical Society.

"I started making phone calls," Mr. Aslin said. "We had both a grass roots campaign with our existing governors and directors, and also a more broad-based community campaign. We also had a specific campaign aimed at the larger companies and foundations in town. We were successful in raising the $8 million dollars." Major donors included Joe and Joyce Hale (for whom Hale Arena is named), the Sutherland family (Sutherland Exhibition Hall is named in honor of the

R. Crosby Kemper, Jr. (Photo courtesy of the American Royal Association).

late Robert R. and Mae G. Sutherland), Mr. and Mrs. Robert W. Wagstaff (for whom Wagstaff Theatre is named) and the I. A. O'Shaughnessy/Wysong Family Foundations, as well as the Hall Family Foundation and the Edith and Harry Darby Foundation. The William T. Kemper Charitable Trust, the Enid and Crosby Kemper Foundation and the Kearney Wornall Charitable Trust provided substantial contributions, as did their trustee, the United Missouri Bank. The W. J. Brace Charitable Trust, the Louetta M. Cowden Trust Fund and the Flarsheim Trust Fund, all managed by Boatmen's Bank, gave generously. Farmland Industries, Inc., the Kansas City Power and Light Company, Kansas City Southern Industries, Inc., the Lester T. Sunderland Foundation and the Sprint Foundation were among many individuals, companies and foundations which gave monetary support to the American Royal. As always, the BOTARs showed commitment to the project, both as an organization and as individual donors. In Mr. Keller's view, "It was definitely a community effort. The people involved in the American Royal acted with vim and vigor."

The project began in 1991 with the "Dance of the Bulldozers" in which American Royal supporter Mary Ellen Purucker and general manager Jim Taylor climbed into the heavy machinery and pulled levers after the operating engineers had moved the dirt. The Governors Exposition Building was renovated to include the American Royal Museum & Visitors Center and the Wagstaff Theatre

The new American Royal. Trail riders, left to right: Mick Aslin, Harry Vold, Roger Shores, and Bill Beaver (Photo reprinted by permission from *The Kansas City Star*, courtesy of the American Royal Association).

and connected to the new areas, Hale Arena and the Central Exhibition Hall.

R. Crosby Kemper, Jr. served as president of the American Royal in 1991-1992. Mr. Kemper's family has been very supportive of the Royal. His grandfather, William Thornton Kemper, is listed as a member of the board of directors in the 1932 American Royal program. Not only did his father, R. Crosby Kemper, Sr., serve as treasurer, but his mother, Enid Kemper, was co-chairman of the BOTARs in 1958. His son, Sandy Kemper, is currently on the board of directors.

"Mick Aslin and I felt we should have an American Royal museum," Mr. Kemper recalled. "The board was not in favor of it, because it was thought that it would be a drain, a money loser. So I underwrote the museum." The museum's holdings include saddles, American Royal pins and memorabilia, Jerry Litton's FFA jacket, Spud Whitman's chaps and an 800-pound skeleton belonging to a hog named Roughneck as well as interactive exhibits.

"When this building was in the concept stage, the education committee worked with the board of directors and the executive committee to create something that really is unique. There is not another museum in the United States connected with a rodeo, livestock and horse show," said Nancy Perry, the Royal's current director of education. Daphne Bitters was the head of the education committee at the time and also served as chairman of the museum planning committee. Sharon Quimby was director of education at that time. The Junior League of Kansas City, Missouri, Inc. sponsored the creation of a docent's manual for the museum. An audio tour of the museum is available; this was a gift from the BOTAR Organization in memory of Forrest L. "Frosty" Langdon II.

Aspects of the 1991 American Royal were documented on film by John Altman. Mr. Altman was commissioned by R. Crosby Kemper, Jr. to make a 25-minute film, which is titled "The American Royal: Tradition of Excellence." The documentary includes archival footage from 1928 and 1948. It is available on videotape and is sold in the museum.

A benefit barbecue, "Boots, Bows and Barbecue" for the new museum and visitors center was held at Hale Arena on Thursday, November 5, 1992. The complex officially opened on Friday, November 6, 1992. "Much of the success of the opening should be credited to Mr. Kemper, who was president at the time, and to his interest in doing it right," said Mr. Aslin, who was president in 1993-1994. Hundreds of volunteers worked on the dedication ceremonies, which included the unveiling of the 16-feet high steel "Bull Wall" sculpture, which was created by former Kansas Citian Robert Morris. According to Stephanie Hoffman Jacobson, who was the

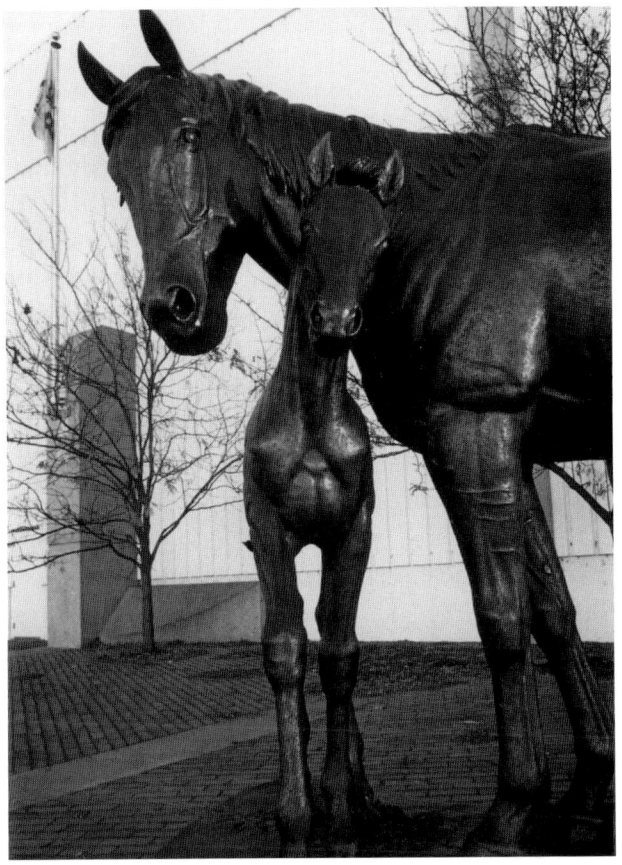

"Spring Days" at the American Royal (Photo courtesy of the American Royal Association).

chairman of the Municipal Art Commission, the "Bull Wall" was the first major "one percent-for-art" project in Kansas City: One percent of the building cost was set aside to pay for the sculpture. The cut-out bulls from "Bull Wall" were placed on "Bull Mountain" at the intersection of Interstate 670, "The Jay B. Dillingham Freeway," and Genessee Street. The Marching Cobras performed. James Madison, president of the Greater Kansas City-Leavenworth chapter of the 9th and 10th Regiments, Horse Cavalry Division, better known as the representatives of the Buffalo Soldiers, carried the American flag in the color guard. Johnie Miller, 93 at the time, rode into the ring for the closing ceremonies. The following day, a "Family Jubilee" featuring jazz and country bands and gospel singers was held at the complex. John D. Free's sculpture of a mare and colt, "Spring Days," is located to the west of the main entrance; the artwork was donated by United Missouri Bank (now UMB Bank, n.a.).

On July 10, 1993, flooding occurred in the West Bottoms, damaging the Royal — especially the newly opened American Royal Museum. Jim Taylor was in South Dakota; he drove all night to get back to the Royal. The lights were out in the building, and everyone carried flashlights to work on the clean-up. As the *Kansas City Star* reported:

> In 1951, Harry Darby sloshed through the flooded American Royal building and vowed that the annual horse and livestock show would return, just as it did after a 1925 inferno... This summer, floodwaters again lapped at Harry Darby's boots. Gold-inlaid footwear that belonged to the late industrialist and politician stood in a display case... A 160-pound silver saddle marinated in 2 feet of inky muck, and a white 1949 Belles of the American Royal gown was stained brown by waters that inundated the Royal

complex.

Restoration of the museum cost more than $200,000; the total damage to the building was estimated at $450,000.

Country music continued to be big business in the 1990s. Many people fondly recall Dolly Parton and Kenny Rogers performing in 1988, and many talented entertainers would grace the stage during the next decade. The American Royal audience which heard Garth Brooks in 1990 was treated to several of the superstar's hits, including "Friends in Low Places," "Unanswered Prayers" and "If Tomorrow Never Comes." The 1994 headliner was the group Diamond Rio. With a nod to the great tradition of country/western song titles, the band performed "This Romeo Ain't Got Julie Yet." Once again, the Royal proved proficient in hiring young talent — Lee Ann Rimes performed in 1996.

Mary Reiff Hunkeler made American Royal history when she became the first woman president, serving in 1995-1996. Mrs. Hunkeler had previously served on the executive committee, the strategic planning committee and the education committee. In 1990, she was chairman of the history committee, which hired James McKinley, the editor of *New Letters* magazine and director of the Professional Writing Program of the University of Missouri — Kansas City, to compile a collection of oral interviews in preparation for the American Royal's centennial.

In November, 1995, the officers, directors and board of governors of the American Royal named Eddie Williams chairman emeritus of the livestock auctions. Less than a month before his death at the age of 83 in December, 1996, Mr. Williams, acting on behalf of Kansas City Southern Industries, bid a record $8,062 for a champion lamb. He raised his own bid from $8000 to $8062 to symbolize his 62-year involvement with the American Royal. Mr. Williams' winning bids totaled more than $3 million to buy livestock at the Royal during those years. He was committed to helping young people fund their education.

It was a blow when the Future Farmers of America, which was organized in Kansas City in 1928, announced in November, 1996, that it would move to Louisville, Kentucky after the 1998 convention. In order to continue the city's commitment to young people interested in farming and livestock, R. Crosby Kemper, Jr. organized the Agriculture Future of America, which awards college scholarships to students from rural areas who intend to work in agribusiness and provides summer internships with many regional corporations and not-for-profit organizations. The AFA has a leadership conference in the fall, and approximately 750 students are expected to attend.

"More than 800 people are taking part in the ceremonies," Anne Turner Wells, chairman of the "Celebration Days" events told the *Kansas City Star* in 1992. "Our mission is to make it clear that the American Royal is for everyone." For many years, the American Royal was seen as appealing to a predominately white audience. In 1991, the fact that two of the young women selected to be BOTARs, Sheryll Lynn Myers and Adrienne Lea Nelms, were African-American made headlines. In reality, Bill Downey's book, *Tom Bass: Black Horseman* created new interest in the African-American heritage of the American Royal from the time it was published in 1975. "A young man by the name of Horace Peterson, who ran the Black Archives of Mid-America, gave me this book and told me to read it," recalled Anthony P. Arnold. "Patricia Jackson, Stinson McClendon, Horace and I started going down to the American Royal and talking to people about Tom Bass." The result of this was the creation of the Tom Bass Reception, which was co-sponsored by the Black Archives of Mid-America Inc. and the American Royal when it was founded in 1991. That year, it was held at the MBL Group Building. The reception has been an annual event at the American Royal since 1992. In addition, the Tom Bass Five-Gaited Missouri-Kansas Stake has been named in honor of the distinguished equestrian.

A noteworthy meeting took place at an early Tom Bass reception when Trooper R. T. Williams,

BOTAR Class of 1991 (Photo courtesy of the BOTAR Organization).

a member of the 9th and 10th Horse Cavalry Association, also known as the Buffalo Soldiers gave his seat to Betty Weldon. The two began talking. The Buffalo Soldiers were working to establish a commemorative stamp, and Mrs. Weldon had experience in such projects and knew ways to expedite the process. The Buffalo Soldiers engaged in what James Madison, first national vice president of the association, has decribed as "a nine-month letter-writing campaign to the Postmaster General." Mrs. Weldon was able to assist by bringing the matter to the attention of Missouri Rep. Ike Skelton. Through the combined efforts of many people, the 29-cent Buffalo Soldiers stamp was issued in 1994.

Tom Bass started the original Tom Bass Riding Clubs as an activity for children. Now the Tom Bass Riders celebrate his memory. "I felt in 1996 that there was such a growth in the African-American community in people riding horses, that we should have a club that was not a club but just an organization. You could stay with your group or be an individual rider, but still ride once or twice a year under the banner of Tom Bass," said Mr. Arnold. The group, the Tom Bass Riders, is a subsidiary of the Twin City Riders, which held its rodeo in Hale Arena for six years.

In 1997, the Tom Bass Riders won second place in the American Royal parade. This is a source of pride for Mr. Arnold. In addition to the American Royal, the group also has gone to the Kentucky Derby. "We were the first African-American riding club to ride in the Pegasus Parade at the Kentucky Derby in 125 years," Mr. Arnold said.

Young people are encouraged to join the Tom Bass riders. "We are so involved with our young people, we review their grade cards," he said. "I'll tell you, it is better than a whipping stick." Teenagers learn responsibility by feeding their horses, taking care of them and maintaining good grades.

Mr. Arnold is the first African-American man, or, as he prefers it "African-American horseman" to serve on the board of directors at the American Royal. (Pam Threatt was the first African-American board member.) He also has served on the committee for the Wild West Show: "When they branded the cattle, I was the one who threw the cattle on the ground."

W. Stinson McClendon, a filmmaker with seven documentaries to his credit, is currently at work on a movie about the life of Tom Bass. At the urging of House Speaker Steve Gaw, Mr. Bass was inducted into the Hall of Famous Missourians at the Capitol in Jefferson City in April, 1999. The bronze bust that bears his likeness joins 19 other prominent individuals from the Show-Me state, including American Royal supporter and United States President Harry S. Truman. As part of the festivities, 160 riders participated in a 45-mile trail ride from the American Saddlehorse Museum in Mexico, Missouri (where Mr. Bass had stables) to Jefferson City.

The Tom Bass Riders at an American Royal Parade (Photo courtesy of Anthony P. Arnold).

1998

Fern Palmer Bittner became the manager of the saddle horse and hunter/jumper shows at the American Royal in 1998. She made her first appearance at the Royal in 1945 when she showed her three-gaited pony, General Ike. The duo received red ribbons that year and the next, and then took the blue ribbon home to the family farm in Columbia, Missouri in 1947.

In April 1998, the Saddle & Sirloin Club announced that it would sell its property at 105th and Mission Road in Leawood and move southeast to a 325-acre site near 139th and Holmes Road in Kansas City. At the beginning of 1999, the Leawood City Council rejected a proposal to develop the Mission Road site, and the club's plans were unclear.

In July, 1998, the *Kansas City Star* reported that a benefit cookout at Walnut Hill Farm, the home of AFA board chairman Sandy Kemper and his wife Christine, "featured grilled steaks from the 1997 American Royal Grand Champion steer, purchased at auction by Crosby Kemper for $40,000 and donated to the AFA."

On Sunday, October 4, 1998, just as volunteers were cleaning up after the American Royal Barbecue, and less than a week before the start of the Arabian show, Kansas City and the American Royal flooded again. The waters rose quickly and unexpectedly in the early evening. "I had worked until about 4:30," Cindy Stanley recalled. "I was filling envelopes with governor's badges and parking passes. And I left them on the floor." Everything had to be redone. But the volunteers and the employees were determined in their efforts, and the American Royal opened on schedule. Cleaning up after the October flood was just one more activity for the approximately 1,000 volunteers who gave more than 12,000 hours of their time in 1998.

Jim Taylor had a surprise in store for his executive assistant. When he casually asked Cindy Stanley what she was planning to do the weekend of the parade, she had no idea she would be riding near the head of the parade in a carriage bearing the sign: Cindy Stanley (heart symbol) of the American Royal. Her grandson, Christopher, rode with her in the carriage, which was donated by Dr. June Miller. Bud Keller was the 1998 Grand Marshal. Bob Hovey did television commentary for the parade, which Bill Harsh used to do. Luanne Neuner was the parade coordinator, a job she has held for many years. Saddle clubs riding in the parade included the Tom Bass Riders, the Hilltop Saddle

Club, the Jackson County Sheriff's Posse, the Rodeo Kids, the Saddle & Sirloin Mounted Patrol and the Wyandotte County 4-H Horse Project.

During the 1998 American Royal, Dr. Hertzog and his colleagues dealt with a potentially disastrous situation. A horse trailer parked outside of Kemper Arena was hit by a bus. "The trailer turned upside down with two horses in it," he recalled. "Fortunately, they were gentle horses. We were able to get in the trailer and tranquilize them. Then the fire department demolished the trailer and got them out. They weren't seriously hurt."

The record for the most money paid for a steer, which was set by Eddie Williams in 1946, held until 1998, when John and Judy Wempe paid $75,000, or $58 per pound for "Big Daddy," raised by 18-year-old Jeff DeRouchey of Pukwana, South Dakota. The Grand Champion Market Lamb was a 152-lb. Suffolk raised by Kami Ott of Fairview, Oklahoma. Kansas City Southern Industries purchased it for $11,000. Elvin Klein bought the Grand Champion Market Hog, a crossbred barrow, which belonged to Jennifer Logston of Royce City, Texas. The hog, which weighed 273 pounds, sold for $10,000. This was a substantial improvement over market prices for swine, which were at their lowest levels since World War II. Many of the animals bought at auction are then donated to charities.

The petting zoo was popular with adults and children alike, but it had one exceptionally busy day: Three sows were in labor at once. Those who could not get close enough to the pens could watch the births on closed-circuit television.

Legendary equitation teacher Helen Crabtree, "the Godmother of Saddle Seat Equitation" and the mother of horse trainer Redd Crabtree, returned to the Royal to be honored at the 50th anniversary of the AHSA Medal. Two riders won the prestigious Triple Crown at the American Royal. Only six other equestrians have ever won. Both riders had been instructed by Lillian Shively and Todd Miles. Andrea Nicole Perry, at 15, became the only rider to ever win the Triple Crown by winning the UPHA walk-and-trot (age 10-and-under), the UPHA junior finals and the UPHA senior finals. Earlier in her career, Ms. Perry had ridden a bay named Reason's Supreme. The horse with whom she won the final leg of the Triple Crown was a chestnut known as Sergeant Royalty. Lauren Murrell and Callaway's Will Gillen (no doubt related to the champion stallion Will Shriver) won the eighth Triple Crown after being named the 1998 AHSA Saddle Seat Medal National Champion. Courtney Sherer and Oh, Shenandoah! placed first in the UPHA junior finals. Miss Sherer's trainers are Carol and Scott Matton.

Longtime Royal competitor Barbara Kirby and her horse, Voyageur, won Class 116, for $300 Amateur Working Hunter, Over 35 Years of Age competition, repeating her 1997 victory. Ms. Kirby and Voyageur also won Class 118, the $300 Amateur Working Hunter.

On the last night of the horse show, the Women's Chamber of Commerce of Greater Kansas City held its annual black tie reception, the Grand Finale, prior to the event. Past activities of the organization have included hosting 4-H breakfasts, sponsoring horse show classes and sponsoring the Grand Champion Junior Divisions for steers, sheep and hogs.

For the FFA convention, American Royal employees wore buttons which said: We (heart symbol) FFA. The buttons were provided by the Kansas City Convention and Visitors Bureau. Many at the Royal expressed regret that the FFA had chosen to break with tradition after 70 years, but there was satisfaction in remembering the many young men and women who had first come to the American Royal as blue jackets and who later returned as adults. Back in 1946, Marion E. Baumgardner of Wellington, Texas, had told the convention that prior to joining the Future Farmers he had never been on a bus, a train or an airplane. As an FFA officer he had flown to Washington D.C. and met the president of the United States. The FFA members of the 1990s are no doubt accustomed to travel — and to virtual travel. The American Royal hopes that they, as individuals, will come back soon and often.

The Centennial

The 1999 American Royal has for the first time national co-chairpersons: Nancy Landon Kassebaum and Howard Baker. The two retired United States senators are married to each other. Senator Kassebaum is no doubt a veteran of Kansas Day. Not only did she serve as a Republican senator from that state, but her father was Governor Alf Landon, the Kansan who was the Republican candidate for president in 1936. In addition, Senator Darby gave young Nancy Landon her first pony. The Royal has selected R. Crosby Kemper, Jr. and Dwight Sutherland, both former presidents, to be honorary co-chairmen. Mr. Kemper, Mr. Sutherland and their families have a long tradition of service to the metropolitan area and especially to the American Royal.

Since 1995, the American Royal has been upgrading its educational programs. The KALF program (Kids Learning Agricultural Fest) has expanded to 11 days in 1999. "Each program centers on a theme, and animals connected to that theme are brought in. For example, during Equine Day, we have 14 different pens of horses for the children to touch and see," said Nancy Perry, director of education. "We have draft horses all the way down to miniature horses. We have activities concerning the day's theme. On Equine Day, we shoe horses. On Sheep Day, we shear sheep. On Grain Day, we plant soybeans, grind wheat and shuck corn. Every day we do an art project that the children can take home." The KALF program runs from April through August; it is targeted at children in sixth grade and younger who are in daycare when school is not in session. On one day in April, 1999, 828 children and 40 volunteers participated in the KALF activities.

The REACh program (Rodeo Education and Children) is held in schools. The 45-minute program teaches children about the different parts of a rodeo. "The children learn about the equipment that a cowboy uses," Ms. Perry said. "It's also an opportunity to show the children that it is important for cowboys to make up their own minds. It's an anti-gang, anti-drug message." In 1998, 8,000 children attended the REACh program. The March, 1999 REACh program was sponsored by the Junior League of Kansas City,

Children at the American Royal (Photo courtesy of the American Royal Association).

Missouri, Incorporated.

In April, 1999, the Royal kicked off its "100 Scholars for 100 Years" program. Each high school in six metropolitan counties (Cass, Clay, Jackson and Platte in Missouri and Johnson and Wyandotte in Kansas) selected three outstanding students, and the area Rotary clubs then chose one student from each school to receive a $5,000 scholarship. Unlike the young people who have brought their animals to American Royal auctions, the Student Ambassadors or the members of the 4-H, the FFA or the AFA, many of these students are not involved in agricultural activities or fields of study. The American Royal simply wanted to thank the community that has given it so much support over the decades.

The BOTARs are celebrating their 50th anniversary. On June 1, 1999, an equine exhibit opened at the American Royal Museum. It is a gift to the museum from the BOTARs. A history of the BOTARs, *Belles of the American Royal Fiftieth Anniversary 1949-1999*, will be published in the fall. Co-chairmen for the book are Amy Taylor Haun and Mary Huxtable Keaveny.

Groundbreaking for the Scott Pavilion, named in honor of American Royal benefactors Betty and Tom Scott, commenced in June, 1999. The pavilion will be used as an exercise area for horses and rodeo animals.

Marianne Kilroy and Dana Hale Nelson are serving as co-chairmen of the Centennial Committee. Events planned for the centennial include a 100-mile trailride from a spot on the Oregon Trail in Kansas through St. Joseph onto the Pony Express Trail and then into the American Royal Complex.

BOTAR past presidents, 1999, left to right: (seated) Marilyn Jurden, Joey Straube, Jo Ann Field, Anne Turner Wells, Margaret Fligg, JoAnn Dickey. (second row) Page Reed, Vicky Leonard, Martha Gail Hughey, Jan Coulson, Nancy Dillingham, Pam McKee, Sharon Franke, Becky Johnson, Julie Richardson, Margaret Hall, Betty Ann Cortelyou, Carolyn Langdon, Joan Kissick, Margaret Freeman, Nancy Thornhill, (third row) Cindy Cowherd, Alison Ward, Karen Turner, Mary Hodge, Julie Fromm, Charleen Fifield, Barbara Ross, Georganne Hall, Sharon Quimby, Marie McMorris, Marianne Kilroy, and Susan Moeller (Photo by Strauss-Peyton, courtesy of the BOTAR Organization).

The Future

When asked in a 1990 interview what he would like to see happening at the American Royal in the year 2000, William Harsh said, "I'd have 17,000 people sitting there for every performance, every one. And I'd like to see the Royal continue to become more important in the education of young people, whether they are raising sheep, painting pictures for the art contest, or being part of Darby's college livestock-judging contests. I'd like to see the people here in Kansas City support that."

Riders on "Bull Mountain" (Photo courtesy of the American Royal Association).

Appendix

American Royal Presidents

1899 - 1902	Kirkland B. Armour
1903	Charles E. Leonard
1904	C. A. Stannard
1905	Allen M. Thompson
1906	George Stevenson, Jr.
1907	Eugene Rust
1908	C. R. Thomas
1909	Nicholas H. Gentry
1910	R. W. Brown
1911	Paul M. Culver
1912	R. H. Hazlett
1913	H. C. Duncan
1914	Charles D. Bellows
1915	Edwin F. Caldwell
1916	R. H. Hazlett
1917	Nicholas H. Gentry
1918 - 1921	James C. Swift
1922	R. H. Hazlett
1923	Charles D. Bellows
1924 - 1925	Edward F. Swinney
1926	John R. Tomson
1927 - 1937	James C. Swift
1938 - 1940	George H. Davis
1941 - 1952	Harry Darby
1953	John B. Gage
1954 - 1955	L. Russell Kelce
1956 - 1957	E. M. Dodds
1958 - 1959	Dallas R. Alderman
1960 - 1961	Jay B. Dillingham
1962 - 1963	Herbert H. Wilson
1964 - 1965	E. K. Hartenbower
1966 - 1967	Roderick Turnbull
1968 - 1969	J. W. Putsch
1970 - 1971	William E. Maurer
1972	John C. Gage
1973 - 1974	Willis C. Theis
1975 - 1976	William P. Harsh
1977 - 1978	Roy A. Edwards
1979 - 1980	Robert D. Hovey
1981 - 1982	Robert W. Wagstaff
1983 - 1984	Dwight D. Sutherland
1985 - 1986	David A. Rismiller
1987 - 1988	Ray R. Evans
1989 - 1990	Charles W. Keller
1991 - 1992	R. Crosby Kemper, Jr.
1993 - 1994	Malcolm M. Aslin
1995 - 1996	Mary Hunkeler
1997 - 1998	Landon Rowland
1999 - 2000	Fred W. Lyons, Jr.

Appendix

Belles of the American Royal Founders

Senator Harry Darby
Donald D. Davis
Dan Fennell, Sr.
Lewis Kitchen
Margaret Weltmer Phinney
(Mrs. Robert Phinney)
Barbara Forrester Rahm
(Mrs. Philip F. Rahm)
F. Forsha Russell
Judy Lupton Woodson Schutz
(Mrs. Carl Schutz)
Miss Nell Snead
Rosemond Richards Straube
(Mrs. Oscar Straube)

Belles of the American Royal Chairmen and Presidents

Chairmen

1949-1953	Barbara Forrester Rahm (Mrs. Philip F. Rahm)
1954-1957	Katherine Stubbs Gambrel (Mrs. Harry M. Gambrel)
1958	
co-chairmen	Enid Jackson Kemper (Mrs. R. Crosby Kemper, Sr.)
	Bernice Rutherford Van Voorst Hanback (Mrs. George Van Voorst; later Mrs. John P. Hanback)

Chairmen and Presidents

1959
chairman   Katharine Buckner Kessinger
(Mrs. Joseph W. Kessinger)
president   Jo Ann Straube Field
(Mrs. Lyman Field)

1960
chairman   Virginia Bee Van Voorst
(Mrs. George Van Voorst III)
president   Anne Turner Wells
(Mrs. J. Lyle Wells, Jr.)

1961
chairman   Eulalie Bartlett Zimmer
(Mrs. Hugh Zimmer)
president   Anne Turner Wells
(Mrs. J. Lyle Wells, Jr.)

1962
chairman   Mary Guinotte Francis

(Mrs. John B. Francis)
president   Marilyn Hudson Jurden

1963
chairman   Wendy Hasek MacLaughlin
(Mrs. William H. MacLaughlin)
president   Marilyn Hudson Jurden

1964
chairman   Sue Schmiederer Luger
(Mrs. George P. Luger)
president   Charleen Dunn Fifield

(Mrs. John R. Fifield)

Presidents

1965	Charleen Dunn Fifield (Mrs. John R. Fifield)
1966	Margaret Campbell Fligg (Mrs. Kenneth I. Fligg, Jr.)
1967	Joey Holter Straube (Mrs. Max Straube)
1968	Olive Beaham Wright Lansburgh (Mrs. Lawrence M. Lansburgh)
1969	Margaret Ann Kurt Freeman (Mrs. Wade J. Freeman)
1970	JoAnn Overman Dickey (Mrs. Gerald L. Dickey)
1971	Karen Van Voorst Turner
1972	Joan Wachter Kissick (Mrs. Robert M. Kissick, Sr.)
1973	Nancy Abbott Dillingham (Mrs. John A. Dillingham)
1974	Charlotte Wornall Kirk (Mrs. Stephen S. Kirk)
1975	Victoria Brigham Leonard (Mrs. George S. Leonard)
1976	Sharon Whitmer Franke (Mrs. Francis S. Franke)

The American Royal: 1899-1999

1977 Susan Bliss Moeller
1978 Georganne Oliver Hall
(Mrs. Porter T. Hall III)
1979 Julie Franz Richardson
1980 Becky Connell Johnson
(Mrs. Eric R. Johnson)
1981 Barbara Smith Ross
1982 Blythe Brigham Launder
(Mrs. David B. Launder)
1983 Julie Evans Fromm
(Mrs. David J. Fromm)
1984 Janelle Wilkerson Coulson
(Mrs. J. Philip Coulson)
1985 Carolyn Robertson Langdon
(Mrs. Forrest L. Langdon II)
1986 Pamela Fogel McKee
(Mrs. Ira William McKee, Jr.)
1987 Susan Cowden Rowan
(Mrs. Roger T. Rowan)
1988 Margaret Weatherly Hall
(Mrs. Thomas B. Hall III)
1989 Betty Ann Cortelyou
1990 Page Branton Reed
(Mrs. Bruce Alan Reed)
1991 Mary Greaves Hodge
(Mrs. Charles Hodge V)
1992 Martha Gail Fogel Hughey
(Mrs. Richard T. Hughey)
1993 Nancy Phillips Thornhill
(Mrs. Thomas E. Thornhill)
1994 Sharon Murray Mueller
(Mrs. Kent E. Mueller)
1995 Sharon Fate Quimby
1996 Marianne Maurin Kilroy
(Mrs. W. Terrence Kilroy)
1997 Cynthia Rapelye Cowherd
1998 Alison Wiedeman Ward
(Mrs. Scott Ward)
1999 A. Marie McMorris

1999 Officers of the Board of Governors

Mary Hunkeler, Chairman
Dwight D. Sutherland, First Vice Chairman
R. Crosby Kemper, Jr., Vice Chairman

1999 American Royal Officers and Directors

Landon Rowland, Chairman
Fred Lyons, President
Perry Sutherland, Vice-President
John Wempe, Treasurer/Corporate Secretary
James D. Taylor, Executive Vice-President /General Manager

Directors

Term Ending January, 2000
John Dillingham
Robert K. Green
Joyce Hale
Mary Hodge
Mary Hunkeler
Jim E. Kay
Alexander C. Kemper
Greg Maday
Fred Merrill
Douglas C. Miller
Dana Hale Nelson
William Rauschelbach
Josh Sosland
Perry Sutherland
John Wempe

Term Ending January 2001
Anthony Arnold
Malcolm M. Aslin
John Berardi
B. Spencer Heddens III
Charles W. Keller
Marianne Kilroy
John Laney
James R. McDowell, Jr.
Paul S. McKie
Edward J. Reardon II
George Shore
Robert E. Smith, DDS
Rod Sturgeon
Thomas S. Ward

Appendix

Anne Turner Wells
Stanley R. Zaremba

Term Ending January, 2002
Barbara Amos
Art Brisbane
Patrick G. Davidson
Fred Ball
Jerry Hedrick
Robert D. Hovey
Jack Kay
Elvin B. Klein
Lee Major
Ira W. "Bill" McKee
William C. Nelson
H. Austin Pollard
Tom Rhone
Dwight Sutherland
Bill Tempel
Larry Wheeler

Additional Voting Members
Bill DuVall
Alison Ward

Honorary Directors
Jay B. Dillingham
Mrs. Roy A. Edwards, Jr.
Mrs. William P. Harsh
Mr. Bill House
Mrs. Robert Wagstaff

Executive Committee
Fred Lyons, President
Malcolm M. Aslin
Robert Hovey
Mary Hunkeler
A. Drue Jennings
Charles W. Keller
R. Crosby Kemper, Jr.
Marianne Kilroy
Paul S. McKie
Marie McMorris
Dana Nelson
William Nelson
Landon Rowland

Tom Scott
Roger Shores
Dwight Sutherland
Perry Sutherland
John Wempe

1998-1999 Board of Governors
(as of July 1, 1999)

Richard Abell
Joe Ablan
Bill Able
Mike Abrams
Mr. & Mrs. Kevin D. Acord
P. James Adam
Dena Adams
Rick Adie
Gary Alexander
Marjorie Alford
Stephen D. Aliber
Kris Allan
A. D. Allen
Barbara P. Allen
Gregory B. Allen
Lance L. Allen
Edwin D. Ammon
Barbara Amos
Betty Amos
Randall J. Anderes
Burke I. Anderson
James R. Anderson
Drs. William A. &
       Sue K. Anderson
Rod Anderson
Stacy Andres
Stacy M. Andreas
Anson Implement, Inc.
Joe Anson
Trisha Anzek
Robert E. Arfsten
Don Armacost, Jr.
Robert S. Armacost, Jr.
Anthony Arnold
Kevin Arnold
James J. Ascher
Evert Asjes, III

The American Royal: 1899-1999

Malcolm Aslin
David Atchley
Tom Atkins
Larry Atzenweiler
David Atzenweiler
Donald & Margaret Austin
Steve Austin
Robert A. Babcock
Ken Bacchus
Nick Badgerow
Teresa Badgerow
Mr. & Mrs. Ron Baker
William R. Baker
Karen Baker
Mark Baldwin
Peggy Schlutius Baldwin
Parnell Baldwin
Mr. & Mrs. Fred Ball
Suzanne M. Ballou
Bill Baragary
Mr. & Mrs. Bruce Barksdale
Doug Barnard
John L. Barnard, M. D.
Ron Barnds
Mark Barrett
James L. Barrick, Jr.
Kevin Barth
Paul D. Bartlett, Jr.
Tom Bartlett
Breck Barton
Thomas Bash
Linda Bastain
Dave & Nancy Baumgartner
John Bennett
James R. Bass Benton
Dave Bastress
W. H. Bates
Ellen J. Wahl & Lee Baty
G. Kenneth Baum
Jonathan Baum
Bob Beagley
Gordon T. Beaham, III
Thomas E. Beal
Jeanne Beals
Bernard J. Beaudoin
Deborah Becker

John W. Beeks, M. D.
Michael Beethe
Howard Behr
Fred Bellemere III
Mr. & Mrs. Shane Belohrad
Tom Bender
John Bennett
William Benson, M. D.
Edward R. Benvenuti, Jr.
John Berardi
Bradley A. Bergman
Bill Bergosh
Doug Bernard
James H. Bernard
James H. Bernard, Jr.
Sara Bernard
Mark Bernhardt
Robert Bernstein
Hon. Robert W. Berrey III
Bob Berry
Tom Berry
Barbara Bierwirth
David Bierwirth
Susan Biggar
Bryan Biggs
Daphne Bitters
Phil Bixby
R. Philip Bixby
Walter E. Bixby
Menefee D. Blackwell
George D. Blackwood, Jr.
Curt Blades
Charles E. Bleakley
Barbara Bollier
Rene Bollier
Carl E. Bolte, Jr.
Stanley Boos
Donald O. Borgman
John A. & Janet W. Borron
Karen Bortz
Robin & Scott Boswell
Mathew Bowen
Gina Bowman-Merrill
Mike Boyce
Peter & Carroll Boylan
Suzanne Bocell Bradley

Appendix

Clint Bradt
Betty Brandes
Joe & Jeanne Brandmeyer
Michael Braude
Rick Brehm
Roxane Brehm
Mr. & Mrs. Jim Brenneman
Douglas M. Briggs
Thomas Brill
Bruce Brinkmeyer
Bill & Wendy Brinton
David Brinton
Reed Brinton
Art Brisbane
Claudia Broaddus
Lonnie Kay Brooks
Paul Broome
John Brown
John O. Brown
Joseph R. Brown
Maynard H. Brown
R. J. Brown
Richard W. Brown
William E. Brown
Hortense Brozman
Loris Brubach, Jr.
Wendall (Pat) Bryan
Jeffrey M. Bublitz
Karl E. Bublitz
Georgia Buchanan
Mr. & Mrs. Matt Buchmann
Ken Buell
Terry Buker
Tom Burcham
Harry Burdg
Kevin Burgoon
Debra A. Burnham
Joseph W. Burns III
Marilynne Bushaw
M. L. Butterworth
Edward E. Butts
John C. Byram, Jr.
Melissa Bynum
Mario Cabrera
Mac F. Cahal
Dorthea Caiazza

Steve Calder
Mr. & Mrs. Harry L. Callahan
Candace Calloway
Karen Calvin
Stephen Campbell
Terry Campbell
Guy R. Cannon
Mel Carnahan
James E. Carnes
Kirk Carpenter
Alice Carrott
Philip W. Carrott
Steve W. Cattron
Rick Chamberlain
Richard K. Champagne
Doug & Terry Champagne
Virginia D. Champagne
Lane Chandler
Dewey Chapman
Wayne C. Chappell
Christy Chester
Robert W. Chester II
Calvin Chestnut
The Rev. Dr. Susan K. B. Chinnery
Meredith Christensen
Jon Christenson
Tara Christiansen
Edward R. Christopherson
Jim Chumley
Jamie M. Clark
Keith Clark
W. E. Clarkson
Emanuel Cleaver II
Harry Cleberg
Candy Clevenger
John N. Clevenger
Kim Clevenger
Steve Clifford
Richard Clinch
John Cline
Clay Coburn
Linda Coburn
Greg Coffey
Jerome Cohen
Roger Coldsnow
Charles R. Cole

The American Royal: 1899-1999

Donald P. Coleman
Dan Colling
Robert Coma
Jeffrey Comment
Al Conway
Dr. David A. Cooley
Ben W. Cooper
Cliff Copp
Cork Corcoran
Gerry Corkle
Carol Costello
David Costello
Merrie Costello
Richard S. Coulson
Cynthia Cowherd
Fran Cox
Jack Crabb
Peggy A. Crabb
Tim Cranor
Mike Crawford
Mr. & Mrs. Samuel Crawford
Dr. Sidney Crawley
Gary Creten
Bill Crooks
Lisa & Roy Crooks
Debbie Crossland
Larry Culleton
Peaches & David Cunningham
Thomas L. Curtis
Scott Cusick
Paul Danaher
William R. Daniels
Edith & Harry Darby Family
Tim Daugherty
Patrick G. Davidson
Thomas J. Davies
Loyd Davis
Raymond F. Davis
Rich & Coleen Davis
Dean Davison
Lester M. Dean, Jr.
Marshall H. Dean
Paul DeBruce
Sharla Decker
Dr. & Mrs. David A. Deer
Steve Dees

Charles J. Defeo III
Mr. & Mrs. Bill Degnan
Jay DeGolar
Art DeMartini
Mark Demetree
Greg DeMint
James C. Denneny, Jr.
Bill Deverill
Vincent & Lisa DeVry
Chester Dibble
Mrs. Gary Dickenson
H. H. Dickenson
Sheila Kemper Dietrich
Jay B. Dillingham
John A. Dillingham
Paul Dinovitz
Beverly Dockhorn
Doug Dockhorn
Dr. Robert J. Dockhorn
John C. Dods
Cathleen Dodson
A. W. Doepke, Jr.
Thomas P. Doherty
Joe Donnelly
Paul Donnelly
George & Edna Dooley
Evan A. Douthit
Charles F. Downey III
Leo P. Dreyer
Robert L. Driscoll
Mark Dudenhoffer
Robert M. Duboc
William J. Duensing, Jr.
Mary Kay Duensing-Lofland
Jerome C. Duggan
Lora Duguid
Stephen A. Dumsky
Bryon Duncan
Anne Dunn
Peggy & Terry Dunn
Robert J. Dunn
Stephen D. Dunn
William H. Dunn
William H. Dunn, Jr.
Bart Dunsford
Mary Nan Dupont

Appendix

Joe & Pat Durkin
Scott Duroe
Daniel S. Durrie
Dan Duvall
William O. Duvall
Nancy Duvall
Tim Dye
John G. Dyer
Joel Ebbert
John S. Eckels
Mimi Eckels
Rochelle Ecker
Greg Edelblute
F. R. Edmunds, Jr.
Joan Edwards
R. A. Edwards III
Eph & Jan Ehly
N. Louise Ellingsworth
Estelle Ellis
Hayne Ellis III
Long Ellis, Jr.
John Ellspermann
Warren Erdman
John Ertz
William C. Esry
Jim Essington
Bev Evans
Edith Marie Evans
Gary Evans
Ray Darby Evans
Ray R. Evans
Robert G. Evans
Denise E. Farris
E. Wayne Farmer
Terri Farwell
Bob Faulkner
Barbara Fay
Jim Bob Feller
J. A. "Jay" Felton
Marcia Fennesy
John H. Ferguson
Judy Ferguson
John Herbert C. Ferney
Louis & Rosemarie
     Fernandez, Jr.
William Dick Fickle

JoAnn S. Field
Laura Kemper Fields
Michael Fields
Ronald Finley
Jerry Fladung
Steve Fletcher
Jeff Flora
Bill Flynn
Janie Flynn
Jerry P. Fogel
Randolph C. Foley
Bryan Folk
Edward F. Ford III
Roger Foreman
Carole Fornelli
Don Forsythe
Leah Foster
Mark Foster
Roy Fowler
Wally Franz
Jerry Freeman
Eugene Freeman
Samuel C. Freitag
Charles French
Norman E. Fretwell
Tom & Sue Frey
David J. Fromm
Dave Fulton
Terry Fulton
John C. Furla II
Hires W. Gage
John B. Gage II
Dick & Jeannie Galamba
Bill Galligan
Dennis Garberg
James C. Garland
James H. Garner, Jr.
Ned Garrigues
Dr. & Mrs. George Gates
Pat Gaunce
Charles Gause
Tim Gelvin
Harriet Gibson
Bill & Janie Gilges
Gayle Gray Gill
Farley Gilliam

Webb R. Gilmore
Larry Glaze
Brian Gloe
Keith Gloe
Jim Glover
Jeff Goble
Wayne Godsey
Dr. John R. Goheen
Mr. & Mrs. Scott J. Goldstein
Charles Goodman
Steve Gordon
Anita Gorman
Gary Goscha
E. Mariese Gourley
Robert J. Gourley
Bill Graham
Bill & Linda Graves
George & Sue Grazier
Great Southern
      Life Insurance Co.
Richard Green
Richard C. Green, Jr.
Robert K. Green
Mrs. Robert C. Greenlease
Fred Ronald Greenstein
Lewis Gregory
Louisa Raich Grill
Bob Grothe
Mario Guastello
Michael R. Gunter
Ronald W. Gurley
Pamela S. Gyllenborg
Kay Bixby Haddad
Dennis Hague
Joe & Joyce Hale
Mr. & Mrs. John Hale
Lisa Hale
Adele Hall
Donald J. Hall
Casey S. Halsey
Debby Hames
Tim Hamill
Gerald Handley
Pete Hannasch
R. Bradley Hansen
Ronald D. Hardten

Kate Hardy
C. W. Haren
Tod Haren
Clifford L. Harris
D. George Harris
Elliott Wendell Harris, Jr.
George Harrison
G. Hall Harsh
Molly Harsh
Mrs. William P. Harsh
Wynona Hartley
Gordon Harton
Ken Hartung
Ron Harvey
Patti Haskell
Todd Hasty
James H. Hathhorn
Mr. & Mrs. Thomas Hauser
Mike Haverty
John E. Hayes, Jr.
Connie Hays & Marlin
      McCutcheon
Marie & Forrest Haynes
Woody Haynes
Diane Hebert
Robert L. Hechler
Dan & Pam Hecker
B. Spencer Heddens III
Chris & Linda Hedemann
Jerry Hedrick
Cindy Heeney
Steven J. Heeney
Mr. & Mrs. Jim Heeter
Lynn Sutherland Heitman
Mr. & Mrs. Barnett C. Helzberg
Tom Henke
Jim Hensley
Bruce Henson
Henry J. Herrmann
Lavon Hess
Greg Hessenflow
Mr. & Mrs. Ken Hickerson
Mrs. John Dible Hickok, Sr.
Sarah B. Higdon
Angela Hill
Hugh F. Hill

Appendix

James R. Hill
Lloyd & Sue Ann Hill
D. W. Hininger
Joy Hise
Robert Hitzhusen
Conrad Hock, Jr.
Jim & Debra Hock
Mary Hodge
Wayne H. Hoecker
John & Linda Hoffman
John L. Hoffman
Sharon Hoffman
George C. Hohl
David Hokanson
Don H. Holder
John Holland
Dave Holtwick
Mr. & Mrs. William J. Honan
Bob Honse
Dan Hood
W. R. Hook
Alvin J. Hooker
Clark Hoover
Dr. Larry A. Hoover
Donald Hopewell
Marc Horner
Mr. & Mrs. Dan Hosfield
Joyce Hoss
Bill House
Carol Hovey
Robert D. Hovey
Joe Howard
John Howell
Mr. & Mrs. Richard Howell
Robert Howell
William D. Howey, Jr.
Katie Hoyt
Douglas L. Hrdlicka
James R. Hudek
Michael S. Hudgeons
Larry B. Huebner
Craig Huffhines
Mr. & Mrs. Stephen R. Hughes
Ray W. Hull
David Hummel
John Hunkeler

Mary Hunkeler
Graham T. Hunt
Trula Guiou Hunt
Erik Hunter
Michael J. Hunter
Sherry Love Hunter
William S. & Sue Hunter
Dean Hurlbut
Carol A. Huseman
Dave Ibbott
David Immenschuh
Barbara Ingram
James Jackson
Joe Jackson
Marcus Jackson
Sally Jackson
Howard T. Jacobson
Larry Janacaro
Mr. & Mrs. A. Drue Jennings
Dick Jensen
Don A. Jenson
Craig Johnson
Cynthia Johnson
Mr. & Mrs. David Johnson
Eric Johnson
Eric & Becky Johnson
Marcie Johnson
Perry E. Johnson, Jr.
Tom Johnson
Mark Johnsrud
Lawson E. Jones
Lowry & Laura Jones
Richard F. Jones
S. Meigs Jones, Jr.
Mary Jane Joyce
Stephen A. Joyce
Dr. Steven T. Joyce
John & Donna Jurco
Ronald Kahn
G. J. Kallos
Hal Kalousek
Larry Kaminsky
Gabrielle Kaniger
Michael & Christine Kaplan
David Kassen
Julia Irene Kauffman

Jack R. Kay
James Kay
Lonnie Kay
Betty Keim
Charles W. Keller
Joseph G. Keller
William A. Keller
Mrs. Claude Kelley
Virginia Kelley
Bruce & Peggy Kelly
Glen Kelly
Joanne Kelly
Laura Kelly
Sandra Kelly
Basil W. Kelsey
Alexander C. Kemper
Christine Kemper
James Kemper
Jonathan Kemper
Mr. & Mrs. R. Crosby Kemper, Jr.
R. Crosby Kemper III
Wm. T. Kemper Foundation
Mark Kenney
William B. Kessinger
J. Lawrence Kilroy
John & Peggi Kilroy
Marianne Kilroy
Terrence Kilroy
Barbara Kirby
Jim Kirk
Mary Francis Kirkpatrick
Steven L. Kitchin
Link Kittrell
Elvin B. Klein
B. Kline
Leonard P. Kline, Jr.
W. Kline
William P. Kline
Earl Knauss
Benjamin O. Knight
Cathy Knight
Dan Knight
H. Elvin Knight
Judith Knight
Susan Knight
Donna Knoell

Charles W. Koester
Carter H. Kokjer
Thomas George Kokoruda
Keith & Kathy Konstantinos
Carl Koupal
Denise Kranzberg
Kipp Kreutzberg
Randy Kreise
Gary Krings
Curtis A. Krizek
Robert Kronschnabel
Anita Krummel
Charles Kruse
Kevin Kurz
Rebekah Kurz
Don Kutz
Dan Ladd
William B. Lague
Sanders R. Lambert, Jr.
Mr. & Mrs. C. Rodney
        LaMothe
Douglas Lancaster
John Landon
John Landry
Debbie Lane
Glenn D. Lane
Marie Lane
Ruby Lane
John Laney
John Lang
Mr. & Mrs. W. W. Lang
A. C. Langworthy, Jr.
Olive Lansburgh
Randy Larsen
Len Larson
Darrell Latham
John Latshaw
Jeanie S. Latz
Brent M. Laue
David Launder
William Lawson
Terry Lay
Mr. & Mrs. Albert Lea
Helen Lea
Gerald & Nancy Lee
Don Legler

Appendix

Ronald T. LeMay
Dr. Tom Lenz
Jackie Lenz
Henry Leonard
Martha Leonard
Peter S. Levi
Richard M. Levin
Anne Gage Lewis
James S. Lewis
Dr. Revis Lewis
Mr. & Mrs. David H. Lillard
Sharon Lindenbaum
Michael Lintecum
Dr. & Mrs. Robert Littlejohn
Teresa Loar
Ann Loder
Dave Lockton
Jeffrey C. Lofland
Bill Loftin
Carol Loftin
Greg Long
Dr. James R. Long
Mr. & Mrs. Mike Long
John Longan
Paula Jo Longan
S. W. Longan
Frank Loschke
Sherry Love
Frank Lowry
Jack Lowry
Bev Lowry-Poplinger
Robert W. Loyd
Jay Lubarsky
Michael A. Luby
Mr. & Mrs. Ray B. Luhnow, Jr.
Jim Lynch
Mr. & Mrs. Fred W. Lyons, Jr.
Lawrence R. Lyons
Steve Lyons
James Maag
Frank Mader
Jack Madill
Lee Major
Lindsay Major
Raymond E. Maloney
David H. Manco

Mr. & Mrs. Phil
    Mangiaracino
Ed Mangone
Carol Marinovich
Alan R. Marsh
Harold & Connie Marshall
Phillip Martens
Thomas O. Martens, Sr.
Becky Martin
David E. Martin
H. Edward Martin, DDS., M.S.
William C. Martucci
Larry M. Marullo
Tom Mathis
Bob Matthews
John W. Matthews
Dave Mattson
A. Ford Maurer
Christine Maurer
Michael T. Mayer
William L. Mayo
Neal McCollum
John N. McConnell
Edward & Linda McConwell
Dr. & Mrs. Stephen McCray
Jenny McDonald
Jim McDonald
John McDonald
Thomas McDonnell
Tom McDonnell
Jeanne McDowell
James R. McDowell, Jr.
Richard E. McEachen
Sally McElwreath
Michael McEntire
John R. McGee
Joseph McGee, Jr.
Simon McGee
Marilyn M. McGilley
Earl & Joan McHugh
Ira W. McKee
Paul S. McKie
Robert W. McKinley
Richard N. McLain
James H. McLarney
Patrick McLarney

Joel McLiney
Edward J. McShane
William McVay III
Roger D. McWherter
JoAnn Meador
Carol Meharry
John Meggs
Charlotte Mehrer
Edward W. Mehrer
Betty Meiners
Steve Melcher
Patrick Melland
Anthony & Connie Mendolia
Mr. & Mrs. Fred Merrill
Fred Merrill, Jr.
Michael A. Merriman
Michael Meyer
Charles R. Meyers
Brett Milbourn
Steve Millard
Bernadette Miller
Craig Miller
Douglas C. Miller
Jane Miller
Dr. & Mrs. Lowell Miller
Merrel Miller
Richard W. Miller
Tim Miller
Ken Millet
Jasper & Lisa Mirabile
Angie Mitchell
John W. Mitchell
Bill Modrcin
James L. Moffett
Mr. & Mrs. Bob Mogren
Lowell Mohler
Joe Monello
Joe Montana
Edward Moody
Larry E. Moore
Tom Moore
Mr. & Mrs. Dale Moorman
Tracy Morris
Mike Morrissey
Mindy Morse
George Morse

Warren Morse
Leo Morton
Loretta Morton
Kim Moshier
Kent E. Mueller
James E. Muiller
Mike Mullin
Lois A. Murdock
Ted A. Murray
Mr. & Mrs. A. H. Myers, Jr.
Charles R. Myers
Marsha Myers
Trula Myers
Roy & Connie Nafziger
Joe & Gloria Nash
Terry Naylor
Mary Williams Neal
Robert Allen Neal II
Dana Nelson
Dr. Douglas L. Nelson
Douglas R. Nelson
Kimi Nelson
William C. Nelson
J. H. Nesselrode
Joseph J. Neuner
Jim Neunuebel
Frank Newby
Frank Newcomer III
Marty & Clyde Nichols
Miller Nichols
Stephen E. & Marianne Noll
F. A. Norden
Cecil Norman
Mrs. Elmer Novaria
Dan Nugent
Steve Nuss
Ned O'Connor
Richard S. O'Neill
Richard S. O'Neill, Jr.
David Odegard
F. A. Olander
David & Mary Oliver
Stephen A. Oliver
Jack Olsen
Diane Olson
Hal Oppenheimer

Lloyd Oppenheimer
Mr. & Mrs. Phil Orscheln
Ronald G. & Nancy R. Osborn
Rosalyn Osborn
Ron Osborne
Kevin C. Ostby
Jane Ott
Jack Ovel
Gene Owen
Richard Owen
Harvey Padewer
Mr. & Mrs. Earl C. Padgett
Douglas F. Page
Paula Palmer
Becky Palmgren
Robert W. Park
Ed Parker
John Parker
Stacey Parker
John & Amy Parry
Jessica Baum Pasmore
Curtis J. Patterson
Linda Patterson
Terri Patterson
Mr. & Mrs. William Patterson
James L. Patton
Sam C. Pearson III
Steve Pearson
Skip Peavey
Kent Pech
Carol Pecoraro
Bill Pederson
Charles Peffer
Margi Pence
Mr. & Mrs. James Pendleton
Ed Peper
James G. Perilstein
Kent Perry
Don Peterson
Tim Petit
Lucinda & Richard E. Petrie
Mike Petrie
Larry Pfeiffer
Natalie Pfost
Ron Pfost
Victor V. Phalen

Beau Pierce
Robert Pierce
Harold W. Pilcher, DVM
Lynn Sutherland Piper
Linna Place
Victor Poirier
Dallas Polen
H. Austin Pollard
James A. Polsinelli
Kathy Polsinelli
Norman Polsky
Mr. & Mrs. Louis Poplinger
Mr. & Mrs. Thomas Pospisil
Mary Ann Powell
Donald H. Pratt
E. Wynn Presson
David Preston
Edward C. Price II
Primedia Intertec
Tom Pruett
Mary Ellen Purucker
Hal Quinn
Kert & Susan Rabe
Richard W. Radke, DDS
Alan R. Raun
Gerald Rauschelbach
William W. Rauschelbach
Randa Rawlins
Christine Raya
Jim Raysik
Mr. & Mrs. Edward J. Reardon II
Ronald & Johyne Reck
Ab Reed
Steve Reed
Travis Reed
Ab Rees
Mr. & Mrs. C. Stephen Reiff
J. Robert & Nora K. Reinhardt
Robert J. Reintjes
William & Victoria Reisler
Wes Remington
Teny Reynolds
Charles E. Rhoades, MD
Tom Rhone
Mr. James O. Riccardi II
Aubrey E. Richardson

The American Royal: 1899-1999

Debra Richmond
Paul Richmond
William M. Richmond
Janet Rickel
Joe Ridgley
Robert E. Riesmeyer
Jerome Riffel
Colonel & Mrs. John Riffle
Norma & Sara Ring
Phil Rinkowski
David A. Rismiller
A. I. Riss
E. S. Riss, Jr.
Edward S. & Jan Riss
J. D. Riss
M. M. Riss
R. R. Riss
Mr. & Mrs. Robert B. Riss
Laura A. Ritchie
Marvin Robertson
Jack Robinson
John H. Robinson, Jr.
Christy Elizabeth Rodes
Ernie Rodina
Leonard C. Rodman
Craig Roeder
Dr. Joseph J. Roh
Ellen Rohde
Randolph Rolf
John & Cynthia Romito
John Rose
Karen Rose
Neil Rose
Chuck Rosley
Mr. & Mrs. David Ross
John J. Ross, Jr.
Pat & Mary Ross
June Rouse
Lawrence Rouse
Landon H. Rowland
Sarah Rowland
Jim Royer
Gwen Royle
John Royle
Dave Ruf, Jr.
Victoria Ruhga

Robert G. Ruisch
Edward Rule
Trey Runnion
Thomas & Nancy Ruzicka
Davis Ryle
Mark J. Sachse
Frank L. Salizzoni
Robert T. Salsman
Ray J. Samuel
Steve Sanders
Robert Sanditz
Wesley J. Sandness
James A. Sangster
Jack & Linda Sauer
Doris Sawyer
Martha Richards Sawyer
Sally Woodson Sawyer
Samuel L. Sawyer
Mario A. Scaglia
Jim & Becky Schaid
Don Schilling
Kevin Schinze
Charles P. Schleicher
Stuart A. Schlemmer
Mark A. Schmidtlein
John Schnakenberg
Steve Schneider
Robert B. Schorb
John Schorie
Gail Shrager
John T. Schroll
David & Ellen Schulte
William P. Schutte
Horst Schwab
Roger H. Schwabauer
Betty C. Scott
Mr. & Mrs. Robert Scott
Mr. & Mrs. Thomas M. Scott
John Scoville
Bill Sears
Richard L. Seithel
John J. Senk, Jr.
Mr. & Mrs. Steve Sestak
Edward A. Setzler
Rick Seymour
Charles N. Sharpe

Appendix

Martha Jo Shaw
Morgan Shay
George Shore
Melissa Shores
Roger A. Shores
Gail Shrager
Donald Shughart
Shughart Thomas & Kilroy
Noel Shull
Burr Sifers
Andrew H. Sigler
Roger & Sharon Sigler
Lloyd T. Silver, Jr.
Al Simmons
Kelvin Simmons
Mark R. Simpson
Curtis & Linda Sims
Ray Sims
Michael & Sunday Siragusa
Mr. & Mrs. Leonard Slaughter
Kevin Sleyster
Skip Sleyster
Trent Slusher
Chris Smart
Missy Smart
Robert L. Smart, Jr.
Cotton Smith
Edwin Smith
Janie Smith
Kit Smith
Larry Smith
Richard Smith
Robert E. Smith, DDS
Shirley K. Smith
Tom E. Smith
Tom & Vickie Smith
Steve Snodgrass
Stephen S. Soden
Scot Soendker
Betsy Solberg
Josh Sosland
Morton I. Sosland
Timothy Sotos
Kelton Spain
Frank & Sandy Spalitto
William M. Spann, Jr.

Mr. & Mrs. Byron Spencer
Mr. & Mrs. Jack Spilker
Monty Spradling
Aggie Stackhaus
Tom & Meg Stafford
Dr. Dwight D. Stanford
Lucinda Stanley
Mr. & Mrs. Dave Stansfield
Tim Stanton
Benjamin Stark
Jim & Alice Stark
Betty E. Starmer
John D. Starr
Jack W. Steadman
Doug Steele
Ron Stein
Darsi Stern
Arthur L. Stern
Alan R. Stetson
Carole Stephens
Richard J. Stephens
Dr. & Mrs. Raymond W. Stockton
Keith Stokes
Clifford Stone
Philip D. Straight
Jan L. Straube
O. Max Straube IV
Phil Strongin
Jim Stuck
Bill Stueck
Otto Stueck
Mr. & Mrs. Charles Stumpf
Rod Sturgeon
Bob Sullivan
Charles & Jackie Sullivan
Jack Sullivan
Tim Sullivan
Matthew R. Sumpter
James P. Sunderland
Paul Sunderland
Robin Sutcliffe
Charles Sutera
Jerry Suther
Pam Suther
Bradford Sutherland
Craig D. Sutherland

The American Royal: 1899-1999

Dwight & Norma Sutherland
Dwight D. Sutherland, Jr.
Herman Sutherland
Jack Sutherland
Mark B. Sutherland
Melody Sutherland
Perry H. Sutherland
Dr. Bill Swafford
Robert Swafford
Dianne Swann
Mike Sweat
Betty Sweeney
Larry Swift
Mr. & Mrs. Steve Swinson
Judy Swope
Mr. & Mrs. M. J. Swords
Louie Swyden
William M. Symon, Jr.
Phil E. Taggart
Robert Takacs
Janet L. Talcott
Betty Talley
Jean & Pete Tamburello
Tanis Family
Bailus M. Tate, Jr.
Karen Taylor
Mary Frances Taylor
William C. Tempel
Ruth Terrill
Bart Thedinger
Mr. & Mrs. Robert O. Thedinger
Stephen O. Theis
Willis C. Theis
Michele Thill
Ken Thomas
Ann J. Thompson
Byron G. Thompson
James C. Thompson
Richard Thompson
Robert M. Thompson
Ron Thompson
Pamela J. Threatt
Jack Tillotson
Mr. & Mrs. John E. Tillotson, Jr.
Nancy Tipton
Mr. & Mrs. Perry Toll

Carl & Mary Tollefson
Clinton K. Tomson
Joseph A. Towns
H. Guyan Townsend III
Liza Rowland Townsend
Clifford R. Trenton
Dennis Triplett
Gary & Kay Truitt
Jim Tucker
Keith & Laura Tucker
Gray Turner
Jill Turner
Prewitt B. Turner
Thomas J. Turner, Jr.
Mr. & Mrs. Thomas J. Turner III
Michael Tutera
Robert & Sally Uhlmann
Mark W. Untersee
Dennis Urban
Joe Uriell
Bruce Van Der Camp
Thomas Van Dyke
Joe H. Vaughan
Judith K. Vogelsang
Busch Voigts
Harry Vold
Dennis Wacknov
Mrs. Harold E. Waddill
Don Wagner
Jim Wagner
John Wagner
Mrs. Robert W. Wagstaff
Melinda A. Waldrop
Robert F. Waldrop
Randall Wallace
Sandy LeForge & Clark Wallace
Tracy Walther
DeAnna Walters
Greg & Sandy Walton
Myron Wang
Scott & Alison Ward
Tom & Debbie Ward
E. Frank Ware
Jay R. Warner
Amanda Warren
Ronald G. Wasson

Appendix

Mr. & Mrs. David Watson
Dorothy Watson
Valerie Watson
Charlie Wear
John & Susan Weaver
Warren Weaver
Al Weber
Mr. & Mrs. Anthony K. Weber
B. L. Webster
Paul Weindenkoph
Stephen Weinberg
John Weisenfels
Paul Weishar
Betty Weldon
Anne Turner Wells
J. Lyle Wells
Mr. & Mrs. James Wells
Russ & Deborah Welsh
John & Judy Wempe
Nancy Wernes
Glenn A. Werry
Kenneth Wescott
Don West, Jr.
Robert H. West
Kevin T. Westrope
Mrs. Harry M. Wheeler, Jr.
Kenneth & Sallie Wheeler
Larry Wheeler
Karl T. Whitacre
James H. Whitaker, MD
John W. Whitaker
Margaret Whitaker
J. Turner White
Richard White
Tom White
James M. Whittier
Kurt Wiedeman
Bruce Wiggins
Robert Wightman
Jim Wilcox
Mr. & Mrs. David Wilkes
Thomas R. Willard
Mr. & Mrs. William E. Willis
Catherine E. Willis-Nugent
John H. Wilson
Don Winter

Joyce Wisecarver
Dr. Ron Witt
David Wittig
Chester & Jeree Wittwer, Jr.
Col. Robert Wolfe
Paula Wolff
Don Wood
Mr. & Mrs. Thomas J. Wood, Jr.
Mr. & Mrs. Mark Woodward
Cheryl Woolery
Larson Woolwine
Tom Woolwine
C Tal & Rosalie Wooten
Mr. & Mrs. Donald E. Wright
Allen P. Wright
Purd B. Wright
Sonya Wright
John C. Wurst
Penny Yates
William E. Yates
Gene Yeager
James J. Yoder
George & Vicki Young
Mr. & Mrs. F. George Zahn
Bernie Zarda
Mr. & Mrs. Stanley Zaremba
James S. Zarr
Tom Zell
John Zimmerman

1998-1999 Honorary Governors
(as of July 1, 1999)

Virginia Holter Alexander
Evert Asjes III
Ken Bacchus
George D. Blackwood, Jr.
Bob Collins
Governor Mel Carnahan
Mayor Emanuel Cleaver
Bill Coons
John E. Cooper
Betty Ann Cortelyou
Janelle Wilkerson Coulson
Cynthia Cowherd
Jack D. Craig

The American Royal: 1899-1999

Paul Danaher
Gerald L. Dickey
Nancy Abbott Dillingham
Ronald Dumay
JoAnn Straube Field
John Fifield
Ronald Finley
Margaret Campbell Fligg
Edward F. Ford III
Mary Harris G. Francis
Julie Fromm
John C. Gage
Jim Glover
Don L. Good
Governor Bill Graves
Georganne Hall
Thomas B. Hall III
C. Coleman Harris
R. E. Hertzog
Charles Hodge
Martha Gail Hughey
Becky Connell Johnson
Marilyn Jurden
Marianne Kilroy
Charlotte Wornall Kirk
Joan Wachter Kissick
Carolyn Langdon
Olive B. Lansburgh
Blythe Brigham Launder
James W. Leathers
Victoria Brigham Leonard
Teresa Loar
Sue Schmiederer Luger
Marsha Giesecke Lundy
Wendy Hasek McLaughlin
Marie McMorris
Mayor Carol Marinovich
Gilbert F. Martin
Pamela Fogel McKee
Barbara Buesking Milledge
Susan Bliss Moeller
George R. Morse
Bobby D. Moser
Mary Williams Neal
Daniel W. Olsen
Sharon Quimby

Page Branton Reed
Julie Franz Richardson
Barbara Smith Ross
Susan Rowan
George Rush
Katheryn J. Shields
Helen Pickering Sifers
Connie Smart
B. C. Snidow
Richard Spader
Aggie Stackhaus
Stanley Stout
Joey Holter Straube
Bill Swicegood
Judith J. Swope
Nancy Thornhill
Ralph Trail
Karen Van Voorst Turner
Virginia Bee Van Voorst
Alison Ward
Jerry Wienberg
Anne Turner Wells
Mary Williams-Neal

1998-1999 Lieutenant Governors
(as of July 1, 1999)

Tony Adams
Jeni Ahring
Mark R. Allen
Paula Amlin
Chad Anderes
Paul & Paul Anselmo
William M. Ashley
Andy Atzenweiler
Thomas P. Barelli
Sherrie Belshe
Barton Bloom
James R. Bocell
Marshall L. Bocell
Jake Bove
Jeff Bowles
David Brinton
Reed Brinton
Anna C. Brous
Matt C. Brown

Appendix

Morgan Brown
Scott Burditt
Niki Burdolski
Joseph Cacioppo
Mr. & Mrs. James R. Carnes
Mary M. Carrott
Geoffrey D &
     Susan Weir Carter
Mary Melissa Chalfant
Claire Nichols Chapman
Cuba Thomas Wagner Chapman
Benjamin C. Chinnery
Hunter R. Christophersen
Anne Elizabeth Cooley
Jennifer Coulson
Clinton Coulter
Heather M. Crabb
Guy W. Creveling
Crude Marketing
John David Cunningham
William M. Dana
Mark L. Davidner
Cy DeVry
John E. DeYoung
Deborah A. Dockhorn
Ryan P. Dolan
Sandra Drake
Diane Drake-Blaylock
Molly J. Dunn
Caroline F. Dupont
Elisabeth Dupont
Nicholas Chapman Dupont
Sarah B. Eckels
Alex E. Eckert
Robert T. Ellis
R. David Emley
John Ertz
Terri Farwell
Jaclyn B. Ferguson
Johnna C. Ferguson
Robb Fleming
Edward Foster
Nancy Freeman
Mark French
Julie Ann Garney
Lisa Michell Garney

John T. Gates
Amy S. Gattermeir
Merlyn & Laura Gibson
William A. Gilges
Louisa Raich & Greg Grill
Anne E. Haines
Alexander C. Hamil
Stephanie Hampel
Laura Marilyn Hanser
Austin Harmon
Barbara Harvey
Danielle Harvey
Melanie Harvey
Joshua Heeney
Nathan Heeney
Andy & Emily Hendricks
Greg Hessenflow
Julie C. Hessenflow
Jon D. Hofer
Sarah L. Hoffman
Amy L. Hunkeler
Britton Hunter
Michelle N. Innes
William M. Jackoboice
Robert L. Jackson III
William Kelley
Christopher Carson Kilroy
Anastasia Z. Knight
Kristin S. Knight
Benjamin O. Knight III
Earl E. Kopp
Brent P. Kroh
Joe LaMothe
Ruby Lane
Mark Lang
Carrie Larson
John Larson
Todd LaSala
Angela Latorre
Joseph Latorre
Christy Anne Littlejohn
Jeffrey C. Lofland
Laura Weber Lutz
Greg & Elizabeth Maday
Catesby A. Major
Robert Manley

The American Royal: 1899-1999

Kimberly A. Mann
Kurt A. Manske
Derek W. Mast
Michael L. Matula
Lindsey McGee
Scott McGee
Peggy S. McGilley
Lindy & Tim McGrath
Margaret McMullen
Christine M. Mehrer
Anthony Michael Mendolia, Jr.
Anthony M. Migliazzo
Kelly Hovey Miller
Michelle P. Miller
Forest Milledge
Holly Milledge
Adam R. Moore
Catherine & David Moore
Julia K. Mulhern
John Nelson
Courtney Lee Nelson
Michael A. Nigro
Mrs. Richard Norden
John Novak
Bridget J. O'Grady
Patricia Louise Owen
Mr. & Mrs. Robert W. Park
Andrew Parker
Blake Parker
Andrew R. Parrott
Kelly Kopp Pasquan
Devon M. Patterson
Evan D. Patterson
Leslie Patterson
Ritchie Anne Patterson
William R. Patterson IV
Chad Perry
Anna L. Petrie
Allison Mae Piper
Preston A. Pollard
Jennifer A. Polsinelli
Marion E. Polsinelli
Patrick Porter
David T. M. Powell
Kent Rainer
Kristin K. Redick

Susan Louise Reiff
Kurt Rhoden
Debra Richmond
Dirk Richter
Michael J. Reilly
Courtney Robbins
Jeffrey Robbins
Kathleen McDowel Robbins
Kelsey Robbins
E. L. Robinson, Jr.
Margaret McGee Ross
William C. Ross
David B. Ruisch
Kim Ruisch
Jennifer Rule
Anne Potter Russ
Jo Marie Scaglia
Philip M. Scaglia
Rebecca Schmidt
John D. Sheridan
Christopher M. Shore
Heather Noel Shore
Scott W. Shore
Sonya G. Shore
Timothy L. Sifers
Christina Silver
Forrest C. Simmons
Courtney Ann Slaughter
Nikki Elizabeth Slaughter
Ryan Slead
Daren Sleyster
Kevin Sleyster
Brandon C. Smith
Lauren Smith
Susan Ambler Spencer
Mary L. Stephens
O. Max Straube IV
Jennifer L. Sullivan
Shelby Sutcliffe
Robert B. Sutherland
John Sykora
Janet Massman Taggart
Carrie Tamburello
Mary & Guy Tamburello
Nicholas Tamburello
Tracie Anne Tempel

Appendix

Charles M. Tetrick
Tracy Lee Lyon Tetrick
Michael R. Thiessen
Barbara W. Thompson
Janet Thompson
Marjorie Thompson
Spencer Turner
Marian & Joe Tutera
Trina Waldrop
Edward F. Walsh IV
Tracy Westlake
Kirsten Kircher Wheatley
Elizabeth A. Whitaker
Dru A. Wilbur
Frank R. Williams
Hayli Williams
Janice Witt
William Wolbach
Larson Woolwine

American Royal Queens

1939	Margaret Jane Swift
1940	Dorothy Jean Ballard
1941	Barbara Montez Dusenberry
1942	—
1943	—
1944	—
1945	—
1946	Connie Daniel
1947	Laura Carol Tarrant
1948	Anabel Baker
1949	Janeice Milrae Bryan
1950	Mary Ellen Ash
1951	Natalie Ruth Kleindienst
1952	Judith Elaine Anderson
1953	De Lois Faulkner
1954	Marlene Hickman
1955	Betty Sue Scott
1956	Mary Jo Smith
1957	Malinda Diggs Berry
1958	Nancy Moore
1959	Mary Diane Arnett
1960	Sarah Kay Burns
1961	Carolyn Jane Parkinson

1962	Jane Riemer
1963	Marilyn Kay "Kay" Hunter
1964	Narka Marie "Mimi" Frink
1965	Deborah Fowler
1966	Evangeline Hope "Eva" Sugarbaker
1967	Elizabeth Ruth Harris
1968	Pamela Anne Brackett
1969	Karen Coffman
1970	Karen Leigh Kemp
1971	Debbie Lee Carey
1972	Lorelie Jean Sousa
1973	Jana Lyn "Jan" Salmans
1974	Shannon Linnell Simonson
1975	Kay Marie Christensen
1976	Holly Stefanyk
1977	Bonita Marie Sneesby
1978	Linda Kaye McGinley
1979	Janet Elaine McNeese
1980	Jan Parcell
1981	Cindy Lou Cooke
1982	Leah Michelle Roberts
1983	Lisa Marie Schaffer
1984	Bernadette Ann Bruening
1985	Geri Lynn Knecht
1986	Belinda Jo Weaver
1987	Shalley Andreen Nottingham
1988	Nicole Sittner

American Royal Student Ambassadors

1989		Bonnie Kay Haws
		Shane Belohrad
	1990	Lisa Killpack
		Stephen Cline
	1991	Jennifer Kapinos
		Thomas Lilja
	1992	Cristy Dicklich
		Larry Whipple
	1993	Melissa Barnes
		Tyler Stuhr
	1994	Sheila Henning
		Jim Barbour
	1995	Sarah Fogleman
		Jerrod Westfahl

1996	Staci Neas
	Matt Schlueter
1997	Sheri Moeller
	Jake Worchester
1998	Andrea Schweitzer
	Matthew Allen Barton

American Royal General Managers

Eugene Rust
C. R. Thomas
Allen M. Thompson
Thomas Wornall
R. J. Kinzer
W. H. Weeks
Frank H. Servatius
A. M. "Andy" Paterson
William E. Preston
C. M. Woodard
George R. Shepherd
Laurence L. Pressly
James "Jim" Taylor

American Royal Horse Show Managers

Allen M. Thompson
Charles W. Green
Edwin C. Eggert
Lon Cox
Bob Leu
J. Ralph Peak
Mary Lou Funderburgh
Marion Vande Wall
Fern Palmer Bittner

American Royal Livestock Managers

A. M. "Andy Pat" Paterson
Dr. Raymond Burns
Arnold Barber
Kenneth Nofftz
Bud Sloan

Saddle & Sirloin Club Founders
(from *History of the Saddle & Sirloin*)

Dan L. Fennell, founding president
E. M. Dodds
L. E. Hawkins
Harry Darby
John B. Gage
R. J. Kinzer
Karl Koerper
W. H. Weeks
E. W. Phelps
Elmer C. Rhoden
Taylor S. Abernathy
Herbert E. Boning, Jr.
J. C. Cash
George W. Catts
C. R. Churchill, Sr.
George Collett
L. P. Cookingham
Donald D. Davis
George H. Davis
Porter T. Hall
W. L. Huggins, Jr.
Senn Lawler
Fred M. Lee
J. Guy Robertson
Perry W. Shrader
Grant Stauffer
Albert H. Wood
Eugene Zachmann

Saddle & Sirloin Club Presidents
(from *History of the Saddle & Sirloin*)

1940-41	D. L. Fennell
1941-42	E. M. Dodds
1942-43	Karl Koerper
1943-44	E. C. Rhoden
1944-45	D. R. Alderman
1945-46	Frank E. Whelan
1946-47	E. F. Phelps
1947-48	George Fiske
1948-49	Frank A. Theis
1949-50	Harry B. Mansfield

Appendix

1950-51	John B. Gage
1951-52	George Van Voorst
1952-53	L. Russell Kelce
1953-54	J. Frank Hudson
1954-55	Herbert H. Wilson
1955-56	Joseph C. Williams
1956-57	E. W. Williams
1957-58	A. K. Simpson, Jr.
1958-59	H. C. Edwards
1959-60	A. D. Eubank
1960-61	John E. Miller
1961-62	Dana Durand
1962-63	S. H. Reno
1963-64	Leo F. Brady, Jr.
1964-65	E. K. Hartenblower
1965-66	Harold E. Purdy
1966-67	Charles F. Monnier
1967-68	J. N. Tiemann
1968-69	Robert S. Armacost, Jr.
1969-70	Ernest B. Hueter
1970-71	Paul A. Tanner
1971-72	F. A. Alexander, II
1972-73	Busch Voigts
1973-74	Robert D. Hovey
1974-75	John A. Young
1975-76	William T. Shields
1976-77	T. C. Llewellyn
1977-78	Lynn V. Bowman
1978-79	Larry C. Nye
1979-80	Stuart L. Murdock
1980-81	Robert E. Smith
1981-82	James S. Lowry
1982-83	Charles W. Keller
1983-84	George R. Morse
1984-85	William L. Lang
1985-86	Alvin J. Hooker
1986-87	Roger A. Shores
1987-88	H. Michael Coburn
1988-89	John E. Wheat
1989-90	Thomas M. Scott
1990-91	Cotton Smith
1991-92	C. W. Haren, Jr.
1992-93	Michael J. Hunter
1993-94	H. Austin Pollard
1994-95	Donald L. Crews
1995-96	G. Donald Winter, Jr.

1996-97	Evan A. Douthit
1997-98	William O. Duvall
1998-99	Lee Baty

AFA Directors
(from the AFA website)

R. Crosby Kemper, Jr.
Alexander C. Kemper
K. Russell Weathers
Michael Braude
Dr. Marc A. Johnson
Dr. Roger Mitchell
Joshua Sosland
Charles P. Schroeder
Frank Sims
Charles Gause
Julian Toney
Timothy R. Dougherty
Gary Hall
Carol Keiser

Top Ten American Royal Volunteers for 1998

Bob Beagley
Thelma Beagley
Irene Endress
Dorie Lamont
Steve Noll
Scot Soendker
Melinda Sparks
Judy Thomas
Keith Thomas
Amanda Warren

1998 Members of the Trail Boss Club
(volunteers with more than 61 service hours for the year)

Burke Anderson
Karen Baker
Merrie Bennett

The American Royal: 1899-1999

Rita Brecheisen
Deb Burnham
Lonnie Kay Brooks
Tara Christiansen
Stephany Conklin
Becky Fitzgerald,
Jean Glessner
John Glessner
Randy Hamm
Roger Hawkins
Robin Hill
Candy Knott
Cathy Lynne
Mendie Melton
Arzella Morrow
Frank Newby
Marianne Noll
Barb Nowling
Jennifer Phillips
Jean Riffle
John Riffle
Irene Shoup
Bob Shoup
Sally Soendker
John Sprugel
Bob Taylor
Mary Thorp
Teresa Tupinio
Tracy Walther
Gina Webb
Ann Welch
Bob Wolfe
Brenda Wolfe
Bob Wolfe
Gene Yeager

1998 Wranglers
(volunteers with more than 41 service hours for
the year)

Slade Baker
Ray France
Robert Gillespie
Rosann Lucas

Marsha Magerl
Vicki Miller
Robbie Moore
Debbie Myers
Tracy Osterkamp
Martin Perez
Christine Raya
Paul Satterfield
Sharon Simpson
Bud Snidow

1998 Scouts
(volunteers with more than 25 service hours for
the year)

Klint Allen
Larry Allen
Terri Becker
Veronica Christiansen
Chris Conklin
John Cooper
Dale Dake
Glenna Dake
Anne Doty
Laura Doty
Vannie Doty
Carol George
Alvino Gibson
Scharlyn Heustis
Sister Mary Laura Huddleston
Janet Hutfles
Becky Jackson
Steve Littlefield
Carol Lopez
Russell McLerran
Rita Miller
Chris Mitchell
Lon Oyster
Jackie Pepper
Whitney Pepper
John Powers
Bea Pray
Lisa Price
Randy Robinson
Karan Sanders

Appendix

Jim Schmidt
John Simma
Joe Towns
Donna Trevolt
Ron Trevolt
Mary Wheeler
Evan Wolfe
Susan Wolfe

American Royal Rodeo Queens

1965	Lacy Giltner
1966	Sharon Kay Colvin
1967	Melissa Newby
1968	Jan Vandeventer

American Royal Swine Showman of the Year

1983	Loyd Kelly
1984	L. W. Tutt
1985	Gale Bressner
1986	Ray Hankes
1987	J. R. Beatty
1988	Charles Peniston
1989	Ray Masters
1990	Deana Guhde
1991	Ron Long
1992	Charles Best
1993	Dan Goehring
1994	
1995	Jim Perry
1996	Don Peter
1997	Jerry Hardin
1998	Leroy Porter

The Star Farmer of America

1929	Carlton Patton, Arkansas
1930	David Johnson, New Jersey
1931	Glen Farrow, Arkansas
1932	Clarence Goldsberry, Missouri
1933	Maurice Dankenbring, Missouri
1934	Paul Astleford, Oregon
1935	Paul Leck, Kansas
1936	Clayton Hackman, Jr., Pennsylvania
1937	Robert Lee Bristow, Virginia
1938	Hunter Roy Greenlaw, Virginia
1939	Norman W. Kruse, Nebraska
1940	Gerald Reyenga, Arkansas
1941	Duane Munter, Nebraska
1942	James H. Thompson, Oregon
1943	Wayne Boothe, Oklahoma
1944	Elton Ellison, Texas
1945	Gordon John Eichhorn, Ohio
1946	William Carlin, Pennsylvania
1947	Ray Gene Cinnamon, Oklahoma
1948	Kenneth L. Cheatham, Illinois
1949	Kenneth England, Arizona
1950	Forrest Davis, Jr., Florida
1951	Harold D. Hodgson, Oklahoma
1952	Walter W. Vogel, Ohio
1953	Stanley A. Chapman, Washington
1954	Burd W. Schantz, Pennsylvania
1955	Joe Moore, Tennessee
1956	Wesley H. Patrick, Georgia
1957	Clarence C. Chappell, Jr., North Carolina
1958	Jimmie John Jarnagin, Jr., Kansas
1959	Lyle Rader, Washington
1960	Arden W. Uhlir, Nebraska
1961	James Isaac Messler, Tennessee
1962	Warner A. Ross, Tennessee
1963	Robert Cummins, New York
1964	Don Carlton Tyler, Pennsylvania
1965	Floyd S. Dubben, Jr., New York
1966	Gary L. Organ, Illinois
1967	David J. Mosher, New York
1968	Joe Boyd Spencer, Oklahoma
1969	Oscar J. Manbeck, Pennsylvania
1970	Merrill Kelsay, Indiana
1971	Lonney Eastvold, Minnesota
1972	David Galley, New York

1973	William A. Sparrow, Georgia
1974	Vernon Louis Rohrscheib, Illinois
1975	Daniel Worcester, Kansas
1976	Timothy H. Amdahl, South Dakota
1977	Dwight Buller, Minnesota
1978	Maynard Augst, Minnesota
1979	Kevin S. Holtzinger, Pennsylvania
1980	Steven Vaughan, Ohio
1981	Chuck Berry, Washington
1982	Kevin Dean Robinson, Kansas
1983	James Tugend, Ohio
1984	Larry Nielson, South Dakota
1985	Mike Arends, Minnesota
1986	Christopher Thompson, Alabama
1987	Franklin Howey, Jr., North Carolina
1988	Clint Oliver, Georgia
1989	Jay Overton, Oklahoma
1990	Todd Lotter, Indiana
1991	Blake Johnson, Nebraska
1992	Kelby Paske, Wisconsin
1993	Daniel Keck, South Dakota
1994	Brian Wade Johnson, Oklahoma
1995	Randy Petroshus, Michigan
1996	Richard Russell, Wyoming
1997	Mike McIntyre, South Dakota
1998	Charles Pearce, Wisconsin

The Star Agribusinessman

1969	Ken Dunagan, Arizona
1970	Earl M. Weaver, Pennsylvania
1971	Wayne Robert Morris, California
1972	Edward D. Higley, Vermont
1973	Steven Redgate, Oklahoma (tie)
1973	Jack Rose, Nevada (tie)
1974	Ronald Dean Schwerdtfeger, Oklahoma
1975	Bryce E. Westlake, Wyoming

1976	Tony V. Pollard, Alabama
1977	Michael Lee Deming, Minnesota
1978	Mark Anthony Williams, Florida
1979	Robert W. Lovelace, Missouri
1980	Jack Baber, Jr., California
1981	Dale Wolf, Jr., Wisconsin
1982	Elmer Zimmerman, Ohio
1983	Clint Albin, Louisiana
1984	Rex Alan Wichert, Oklahoma
1985	Scott Cochran, Georgia
1986	Todd Wilkinson, Tennessee
1987	Dan Ruehling, Minnesota
1988	Christopher Bledsoe, Missouri
1989	David Tometich, Iowa
1990	Chad Luthro, Iowa
1991	Adam Schumacher, Minnesota
1992	Chad Wells, Tennessee
1993	Jeffrey Cole, Missouri
1994	David Snyder, New York
1995	Wade Kallevig, Minnesota
1996	Chad Bischoff, Michigan
1997	Mark Dudgeon, Ohio
1998	Andrew Tygrett, Iowa

National FFA Presidents

1928-29	Leslie Applegate, New Jersey
1929-30	Wade Turner, North Carolina
1930-31	Leslie Fry, Missouri
1931-32	Kenneth Pettibone, Oregon
1932-33	Vernon Howell, Oklahoma
1933-34	Bobby Jones, Ohio
1934-35	Andrew Sundstrom, South Dakota
1935-36	William Shaffer, Virginia
1936-37	Joe Black, Wyoming
1937-38	J. Lester Poucher, Florida
1938-39	Robert Elwell, Maine
1939-40	Ivan Kindschi, Wisconsin
1940-41	D. Harold Prichard, Mississippi
1941-42	Irvin J. Schenk, Indiana
1942-43	Harold Gum, West Virginia
1942-43	Marvin Jagels, Idaho

Appendix

1943-44	Robert Bowman, California		1987-88	Kelli Evans, Nebraska
1944-45	Oliver H. Kinzie, Oklahoma		1988-89	Dana Soukup, Nebraska
1945-46	J. Glyndon Stuff, Illinois		1989-90	Donnell Brown, Texas
1946-47	Gus R. Douglass, Jr., West Virginia		1990-91	Mark Timm, Indiana
			1991-92	Lee Thurber, Nebraska
1947-48	Ervin Martin, Indiana		1992-93	Travis D. Park, Indiana
1948-49	Doyle Conner, Florida		1993-94	Curtis D. Childers, Texas
1949-50	George Lewis, Illinois		1994-95	Corey D. Flournoy, Illinois
1950-51	Walter Cummings, Oklahoma		1995-96	Seth Derner, Nebraska
1951-52	Donald Staheli, Utah		1996-97	Corey Rosenbusch, Texas
1952-53	Jimmy Dillon, Louisiana		1997-98	Hillary Smith, Georgia
1953-54	David H. Boyne, Michigan		1998-99	Lisa Ahrens, Iowa
1954-55	William D. Gunter, Jr., Florida			
1955-56	Daniel Dunham, Oregon			
1956-57	John M. Haid, Jr., Arkansas			
1957-58	Howard Downing, Kentucky			
1958-59	Adin Hester, Oregon			
1959-60	Jim Thomas, Georgia			
1960-61	Lyle Carpenter, Colorado			
1961-62	Victor Butler, Jr., Florida			
1962-63	Kenny McMillan, Illinois			
1963-64	Nels Ackerson, Indiana			
1964-65	Kenneth Kennedy, Kentucky			
1965-66	Howard Williams, North Carolina			
1966-67	Gary Swan, New York			
1967-68	Greg Bamford, Colorado			
1968-69	Jeff Hanlon, Oregon			
1969-70	Harry Birdwell, Oklahoma			
1970-71	Dan Lehmann, Illinois			
1971-72	Tim J. Burke, Iowa			
1972-73	Dwight O. Seegmiller, Iowa			
1973-74	G. Mark Mayfield, Kansas			
1974-75	Alpha E. Trivette, Virginia			
1975-76	Bobby Tucker, Texas			
1976-77	C. James Bode, Jr., Oklahoma			
1977-78	J. Ken Johnson, Texas			
1978-79	Mark Sanborn, Ohio			
1979-80	Douglas C. Rinker, Virginia			
1980-81	Mark Herndon, Oklahoma			
1981-82	Scott Neasham, Iowa			
1982-83	Jan Eberly, California			
1983-84	Ron Wineinger, Kansas			
1984-85	Steve Meredith, Kentucky			
1985-86	Rick Malir, Kansas			
1986-87	Kevin Eblen, Iowa			

Bibliography

American Royal. [newsletter]. Spring, 1999.

The American Royal Encyclopedia.

American Royal Live Stock Premium List 1931.

The American Royal Museum & Visitors Center Docent Manual. Sponsored by the Junior League.

The American Royal: A Special Presentation on Its History, Activities and Plans for the Future [Press release], 1979.

American Royal: Tradition of Excellence [Film]. Dir. John Altman. Masters Mark, 1992.

Angel, Traci. "Missouri salutes Tom Bass." *Kansas City Star.* Tuesday, April 13, 1999, p. B-3.

Barr, Paula. "Royal Crowds No Longer Have Kickin' Joe To Cheer Around." *Kansas City Star* [On-line], Monday, November 16, 1992, p. B-2.

Barr, Paula. "Saddle Up! Riding clubs carry on the African- American tradition of horsemanship." *Kansas City Star Magazine* [On-line], Sunday, January 26, 1997.

Boyle, Robert H. and Alice Higgins. "End of a Bloody Bad Show." *Sports Illustrated,* June 4, 1973, p. 36.

Borman, Dawn. "Horse club's move thrown into limbo." *Kansas City Star.* Wednesday, January 6, 1999, p. B-4.

Breeder's Gazette. October 24, 1900.

Brisbane, Art. "Ray Davis' life was stockyards. *Kansas City Star* [On-line], May 6, 1991.

Bibliography

Bulger, William. "Veteran Royal Follower Misses One." [obituary for John Haffey] *Kansas City Star*, 1966.

Butler, Robert W. "Arena in Peril: Bonds Refused." *Kansas City Times*, Saturday, January 20, 1973, p. 1-A.

Butler, Robert W. "100 Years of Royal images: Filmmaker John Altman sought still photos of city's past for new documentary-making debut. *Kansas City Star* [On-line], November 6, 1992.

Buttry, Stephen. "Kemper Cites Enthusiasm about American Royal." *Kansas City Star*, January 25, 1991, p. C-1.

The Call (Kansas City). "Anthony P. Arnold Named To American Royal Board." Vol. 78, no. 29. February 6-12, 1998.

Carroll, Robert L. "Arena in Peril: Council Revolt." *Kansas City Times*. Saturday, January 20, 1973, p. 1-A.

Carroll, Robert L. "Arena O.K. Expected as Pressures Rise." *Kansas City Times*, Monday, January 22, 1973, p. 1.

Carroll, Robert L. "Council Votes Bonds For Royal's Arena." *Kansas City Times*, Tuesday, January 23, 1973, p. 1-A.

Collins, John M. "Premium Prices Feature the Royal's Largest Junior Show." *Weekly Kansas City Star*, Wednesday, October 17, 1945, p. 1.

Combs, Loula Long. *My Revelation*. Lee's Summit, Missouri: Longview, 1947 (1991 reprint).

Cooper, Brad. "Saddle & Sirloin agrees to sell Leawood facility." *Kansas City Star*, April 23, 1998, p. 1.

Crayton, Rasheeda. "Missouri, Oklahoma students win Royal Ambassador awards." *Kansas City Star*, November 14, 1998, p. C-2.

Cruz, Humberto. "Rousing Start for Royal Rodeo." *Kansas City Star*, June, 1967.

Daily Drovers Telegram. "One Thousand Strong." Tuesday, November 9, 1926, p. 6.

Daily Drovers Telegram. "Program For Royal Week." Tuesday, November 9, 1926, p. 6.

Dauner, John T. "Royal Rule Stirs Up Controversy. *Kansas City Times*, October 23, 1969.

Dauner, John T. "Intrigue and Death Shatter Wealthy Brothers' Lives." *Kansas City Times*, June 6, 1971, p. 1.

Dauner, John T. "Chestnut is on top as grand champion of harness horses." *Kansas City Times*, Wednesday, November 9, 1983, p. B-8.

Dauner, John T. "Night of Roses finds evening of victory." *Kansas City Times*, Thursday, November 10, 1983, p. B-8.

Dauner, John T. "Black stallion strides off with grand championship." *Kansas City Times*, Friday, November 11, 1983, p. B-8.

Downey, Bill. *Tom Bass: Black Horseman.* St. Louis, Missouri: Saddle and Bridle, 1975.

"Facts, Figures & History, American Royal Livestock, Horse Show & Rodeo." [internally published memo, c. 1979-80].

Fisher, James J. "Royal to Buy Its Own Site in Stockyards." *Kansas City Times*, Thursday, December 28, 1972, p. 1-A.

Fletcher, Baylis John. *Up the Trail in '79.* Norman: The University of Oklahoma Press, 1968.

Flexner, Stuart Berg and Lenore Crary Hauck, eds. *The Random House Dictionary of the English Language* (2nd ed). New York: Random House, 1987.

Fowler, Dick. *Leaders in Our Town.* Kansas City: Burd & Fletcher, [1952].

Franklin, William B. "Of Round-Ups and Rocking Chairs." The Kansas City Magazine. October, 1973, p. 26.

Fussell, James A. "The buck starts here." *Kansas City Star Magazine*, Sunday, October 25, 1992.

Garbus, Kelly. "Stallion captures 5-gaited honors." *Kansas City Star*, Sunday, November 24, 1996, p. B-1.

Garbus, Kelly. "'Mr. American Royal,' Eddie Williams, dies." *Kansas City Star*, December 5, 1996, p. 1.

Garden City [Kansas] *Telegram.* "Kansas girl's hog wins big." December 13, 1984.

Gray, Melanie. "That's no bull! This American Royal Guide will teach you to speak with authority on more than just livestock." *Kansas City Star Magazine* [On-Line]. Sunday, October 27, 1996.

Hale, Chief George C., editor and compiler. "Facts About Kansas City, U.S.A." Souvenir of Kansas City and Her Fire Department to the Grand Paris Congress and Exposition. Paris, France. August 13, 1900. Kansas City: Hailman-Reily, 1900.

Haskell, Henry C., Jr., and Richard B. Fowler. *City of the Future.* Kansas City: Frank Glenn, 1950.

Hazelton, Jno. M. "American Royal A Kansas City Institution." *Hereford Journal*, November 1,

Bibliography

1933.

Helliker, Kevin. "Why New York Is Now Outstripping Kansas City Steaks." *Wall Street Journal.* Monday, July 6, 1998, p. 1.

Hendricks, Mike and Joseph Rebello. "Another crisis, another show." *Kansas City Star.* Sunday, October 31, 1993, p. A-1.

Hereford Journal. "End Unexpectedly for R. J. Kinzer." September 15, 1952, p. 24.

Hereford Journal. "Bud Snidow Retires From AHA Duties." June, 1983, p. 30.

Hockaday, Laura Rollins. "Waltzing into history: Cultures and talents blend in first integrated BOTAR class. *Kansas City Star* [On-Line], October 27, 1991.

Hockaday, Laura Rollins. "So big it takes three days to celebrate: New American Royal Complex is focus of ceremonies, entertainment." *Kansas City Star* [On-line], November 4, 1992.

Hockaday, Laura Rollins. "Flood of '93 brings out memories for Royal veteran." *Kansas City Star*, Sunday, November 7, 1993, p. H-4.

Hockaday, Laura Rollins. "AFA fund-raiser at Walnut Hill Farm." *Kansas City Star*, July 19, 1998, p. I-7.

Hockaday, Laura Rollins. "Royalfest creates fun, funds for AFA interns." *Kansas City Star*, August 16, 1998, p. H-7.

Hollan, William. "Jumping event winner gets his Irish up." *Kansas City Times*, Saturday, November 19, 1988, p. C-7.

Hollan, William. "Virgin Islander Wins Grand Prix Rusty Holzer, Riding Picasso, Takes $25,000 In 10-Horse Jump-off." *Kansas City Star* [On-Line], Sunday, November 21, 1992, p. C-8.

Hollan, William. "Texan Triumphs In Contest Despite Hectic Schedule Woman Splits Time Between Oklahoma and American Royal Shows." *Kansas City Star* [On-Line], Saturday, November 20, 1993, p. C-4.

Hughes, Wilma. "Champagne Fizz and the Hulse Magic." *National Horseman.*

Hughes, Wilma. "A Tribute to One of the Greatest — Sug Utz!" *National Horseman.* November, 1987.

Hughes, Wilma. "Lee and Jane Fahey." *National Horseman.*

Hughes, Wilma. "Bill Harsh — 'Mr. American Royal' Remembered with Love And Affection." *National Horseman.*

Hughes, Wilma. "Kansas City and Saddlehorse Industry Say Farewell to H. W. 'Buck' Hinson." *National Horseman*, February, 1998, p. 127.

Igoe, Ruth E. "American Royal adds to scholarship program." *Kansas City Star*, Thursday, April 8, 1999.

The Independent. November 17, 1923.

The Independent. November 20, 1926.

The Independent. "Highlights of the Elsa Maxwell Ball." October   21, 1939, p. 11.

The Independent. October 19, 1946, p. 5.

The Independent. October 23, 1948, p. 10.

The Independent. "Over My Shoulder." Saturday, October 17, 1953, p. 14.

"J. Ralph Peak." American Royal [press release], 1978.

Johnson, Roger T. "Kansas City Board of Trade" in *At the River's Bend: An Illustrated History of Kansas City, Independence and Jackson County* by Sherry Lamb Schirmer and Richard D. McKinzie. Produced in association with the Jackson County Historical Society. Woodland Hills, California: Windsor, 1982.

Jones, Linda Newcom. *The Longview We Remember*. n.p.: Storm Ridge, 1990.

Jones, Robert E. "A Short History of the American Royal."   American Royal [Press release] c. 1966.

Kansas City Board of Trade. Historical Timeline. [On-line].

Kansas City Star. Nebraska Clothing Co. [Advertisement]. Thursday, October 26, 1899, p. 10.

Kansas City Star. Emery, Bird, Thayer & Co. [Advertisement]. Friday, October 27, 1899, p. 8.

Kansas City Star. "Marconi's Navy Experiments." Friday, October 27, 1899, p 9.

Kansas City Star. The Palace Clothing Co. advertisement. Friday, October 27, 1899, p 9.

Kansas City Star. "What The Market Offers." Friday, October 27, 1899, p 9.

Kansas City Star. "The Queen of the Horses." Saturday, October 28,1899, p. 7.

Kansas City Star. "Dale Gets the Armour Cup." Saturday, October 28, 1899, p. 1.

Kansas City Star. "Royal Building the Largest." November 19, 1922.

Bibliography

Kansas City Star. "Lindsborg chorus sings 'Messiah.'" November 19, 1922.

Kansas City Star. "Horses Step High." November 17, 1929.

Kansas City Star. "A Brilliant Royal." November 19, 1929.

Kansas City Star. "Sweetheart Again." October 27, 1935.

Kansas City Star. "It's Our Royal." Sunday May 8, 1938, p. 1.

Kansas City Star. "$150,000 to Woolf." Sunday May 8, 1938, p. 1.

Kansas City Star. "Kansas Day Musicians Examine The 'Tops' — In Kansas Mules." Monday, October 17, 1938, p. 1.

Kansas City Star. "Events of Other Years." Sunday, October 20, 1946.

Kansas City Star. "Royal Sets a Record." Sunday, October 27, 1946, p. 1.

Kansas City Star. "First To Wichita Girl." Sunday, October 27, 1946, p. 2-A.

Kansas City Star. "Arena to Mules." Wednesday, October 22, 1947, p. 1.

Kansas City Star. "Supreme in Ring." Sunday, October 23, 1949, p. 1.

Kansas City Star. "Brother Killed in World War II Motivates Girl's Royal Hopes." Monday, October 16, 1950, p. 10.

Kansas City Star. "Replica A Victor." Sunday, October 22, 1950, p. 1.

Kansas City Star. "Disaster as The Kaw Spreads." Friday, July 13, 1951, p. 1.

Kansas City Star. "Cup To Replica." Sunday, October 21, 1951, p. 1.

Kansas City Star. "Top Show Horse." Sunday, October 26, 1952, p. 1-A.

Kansas City Star. "Royal Show Is Ready." October, 1954.

Kansas City Star. "Trophy To Mare." Sunday, October 24, 1954, p. 1-A.

Kansas City Star. "An unannounced Eddie Fisher appearance..." [photo caption] Thursday, October 13, 1955, p. 8.

Kansas City Star. "Top Steers Are Sold." Wednesday, October 22, 1958, p. 1.

Kansas City Star. "A Bonus For The Royal." Wednesday, October 22, 1958, p. 6-A.

Kansas City Star. "An Angus Wins A Royal Crown." Tuesday, October 17, 1961, p. 1-A.

Kansas City Star. "Champ Brings $4,998." Wednesday, October 18, 1961, p. 1-A.

Kansas City Star. "Royal Honors Oklahoma Girl." October 17, 1962, p 1.

Kansas City Star. "Coronation Tops Royal Ball." October 10, 1965, p. 1-A.

Kansas City Star. "Mrs. Loula Long Combs, Horsewoman, Dies at 90." Tuesday, July 6, 1971, p. 1.

Kansas City Star. "Bulletin." Monday, January 22, 1973, p. 1.

Kansas City Star. "City Hall Makes the Big Decision To Build Sports Arena" [editorial]. Wednesday, January 24, 1973, p. 44.

Kansas City Star. "New American Royal Complex will be dedicated on Friday." November 4, 1992. [On-line].

Kansas City Star. "Edwin Wade Williams" [obituary]. Thursday, December 5, 1996, p. C-4.

Kansas City Star. Obituaries. [Jane Fairchild Fahey]. July 11, 1997. [On-line].

Kansas City Star. "Horseman Tom Bass deserves recognition." Sunday, April 18, 1999.

Kansas City Times. "Notes Of The Big Show." Tuesday, October 13, 1908, p. 2.

Kansas City Times. "Tribute To W. A. Cochel." Thursday, October 24, 1946, p. 3.

Kansas City Times. "A Fine Testimonial..." [photo caption: Loula Long Combs] Thursday, October 24, 1946, p. 3.

Kansas City Times. "Color In Arena." Friday, October 25, 1946, p. 1.

Kansas City Times. "A Rebel Rocks F.F.A." Friday, October 25, 1946, p. 4.

Kansas City Times. "Lively At Royal." Tuesday, October 16, 1951, p. 1.

Kansas City Times. "At Royal Helm." April 10, 1954.

Kansas City Times. "Thrills in Arena." Wednesday, October 20, 1954, p. 1.

Kansas City Times. "Kansas Beauty Is Royal Queen." October 17, 1959, p. 1, 7.

Kansas City Times. "Tom Bass, 'Father of Royal Horse Show.'" Friday, October 23, 1959.

Kansas City Times. "Tables Turned at Horse Show." October 20, 1965.

Bibliography

Kansas City Times. "Stallion Wins 5-Gaited Stake." Wednesday, October 18, 1967, p. 1-A, 6-A.

Kansas City Times. "Services Set For Mrs. Durand." Tuesday, September 29, 1970, p. 2-B.

Kansas City Times. "The Great Lady of Longview." Wednesday, July 7, 1971, p. 36.

Kansas City Times. "R. Crosby Kemper, Sr., Banker, Dies." Wednesday, October 25, 1972, p. 1-A.

Kansas City Times. "Jerry Litton, Family Die in Plane Crash; Final Tally Shows He Won Senate Bid." Wednesday, August 4, 1976, p. 1.

Kansas City Times. "Kansas City Is Down to the Wire On the Sports Arena" [editorial]. Friday, January 19, 1973, p. 32.

Kaye, Joseph. "Easy Way With the Unusual a Method of Dave Galloway." *Kansas City Star*, Sunday, October 16, 1955.

Kemper, R. Crosby, Jr. *Maverick from the Herd.* Privately printed, n.d.

Kozlowski, Kim. "Riding out of history: Charro brings a Mexican tradition to the American Royal. *Kansas City Star* [On-line], November 19, 1998.

LaFave, Kenneth. "Symphony matches the beat to rhythm of cantering horses." *Kansas City Star.* November 11, 1988.

Laird, Landon. "Talk of the Town." *Kansas City Star.* October 13, 1966.

Leathers, James W. and Dillingham, Jay B. "Yards Is Home of Kansas City Steak." Promotion, circa mid-1960s.

Lee, Amy Freeman. *Hobby Horses.* New York: Derrydale Press, 1940.

Lind, Bette. "Silver cup back home after 82-year journey." *Kansas City Star.* May 5, 1981.

Lind, Bette. "Heroes of the pre-auction action." *Kansas City Star.* Sunday, November 8, 1981, p. 20-F.

Lundy, Liz. "The Great American Royal." *Horse World.* November 1953, p. 15.

McCorkle, William L. "Mayor Plans to Move On Stockyards Arena Site." *Kansas City Star.* Monday, January 22, 1973, p. 17.

McCorkle, William L. "Two Obstacles on Arena Pact." *Kansas City Star.* Friday, January 19, 1973, p. 3.

McGuff, Joe. "Station May Be Arena Rallying Point." *Kansas City Star*, c. December 1972.

Miller, Lynda Ann. "Matrons of Distinction." *National Horseman*, May, 1975, p. 98.

Montgomery, Rick. "Spirit of the Stockyards." *Kansas City Star Magazine* [On-line], Sunday, October 25, 1992.

Moore, Bill. "American Royal's First Queen Was School Children's Candidate." *Kansas City Star*, October, 1967.

National Association of State Universities and Land-Grant Colleges. "The Land-Grant Tradition." [On-line].

1996 American Royal Horse Show Program.

1998 American Royal Horse Show Program.

Norton, Bill. "The Men Who Saved The Royal." *Kansas City Star Magazine*. Sunday, October 28, 1984, p. 15.

"100 Scholars for 100 Years." American Royal press conference pamphlet, April 7, 1999.

Ornduff, Donald R. *The First 49*. Introduction by B. C. Snidow. Kansas City: Lowell Press, 1981.

Patterson, Kathleen and John A. Dvorak, "Craft Down On Takeoff; All Dead." *Kansas City Times*. Wednesday, August 4, 1976, p. 1.

Paxton, Heather N. "The American Royal." Script for audio tape for the American Royal Museum. 1997.

Paxton, Heather N. *The Kansas City Country Club Centennial Book*. Kansas City, Missouri: privately printed, 1996.

Pearman, Robert. "Missouri Mule Back in Triumph." *Kansas City Times*. Wednesday, October 16, 1963, p. 1.

Peterson, Ann. "American Royal gains attendance despite decline this year at rodeo." *Kansas City Star*. Sunday, November 20, 1988, p. 14-B.

Pfeffer, Joseph G. "The Saddlebred in Cowtown: Kansas City and the American Royal." *The American Saddlebred*, January/February 1991, p. 70.

Popper, Joe. "Movie could help restore horseman's prominence." *Kansas City Star* [On-line], Saturday, November 14, 1998.

Ralls, Richard D. "Crystal Gayle's Success Transcends 'Brown Eyes.'" *Kansas City Times*, Saturday, November 12, 1977, p.7-B.

Revsim, Michael J. "Family Tradition Helps Him Win Star Farmer Award." *Kansas City Times*,

Bibliography

Friday, November 10, 1978, p. 1-B.

Richardson, Jolie. "The ASHA Saddle Seat Medal National Finals." *National Horseman*, January, 1999, p. 96.

Richardson, Jolie. "The UPHA Junior Challenge Cup National Finals." *National Horseman*, January, 1999, p. 114.

Richardson Jolie and Christy Judd O'Donnell. "1998 United Professional Horseman's Association and American Hackney Horse Convention." *National Horseman*, February, 1998, p. 65.

Roberts, Roy. "A Future Aglow." *Kansas City Star*, Friday, October 25, 1946, p. 1.

Rogers, Will. "Will Rogers Says Kansas City Necks Sore From High Collars." November 28, 1928. Reprint.

Runnion, Dale F. *The Saddle and Sirloin Portrait Collection.* Louisville, Kentucky: Gateway Press, 1998.

Sanders, Alvin H. *The Story of the Herefords.* Chicago: Breeder's Gazette [Sanders Publishing], 1914.

Sandy, Wilda. *Here Lies Kansas City: A Collection of Our City's Notables and Their Final Resting Places.* Kansas City, Missouri: Bennett Schneider, 1984.

Savary, Erica. "Equitation at its Best." National Horseman, January, 1999, p. 126.

Schirmer, Sherry Lamb and Richard D. McKinzie. *At the River's Bend: An Illustrated History of Kansas City, Independence and Jackson County.* Produced in association with the Jackson County Historical Society. Woodland Hills, California: Windsor, 1982.

Scrapbooks belonging to the American Royal.

Scrapbooks belonging to R. Crosby Kemper, Jr.

Scrapbooks belonging to Gladys Mackey.

Scrapbooks belonging to Melissa and Roger Shores.

Scrapbooks belonging to Hilda and Bud Snidow.

Scrapbooks belonging to Sue and George Zahn.

Snidow, Bud. *50 Years of Kansas Ranching: CK Ranch.* n.p.: privately printed, 1984.

Sullinger, Jim. "Buyers break records for Royal's auction." *Kansas City Star*, Sunday, November 15, 1998, p. B-4.

Symons, Bill and Sue. *Saddle & Sirloin: The First 50 Years*. Introduction by David C. Alford. n.p.: n.p., 1990.

Tenney, Dr. A. Webster. *The FFA at 50*. Alexandria, Virginia: Future Farmers of America, 1978.

Thompson, Thomas. *Blood and Money*. Garden City, New York: Doubleday, 1976.

Tomson, Clinton K. Letter to Bud Snidow. May 12, 1999.

Trillin, Calvin. *American Fried*. New York: Doubleday, 1974.

Trillin, Calvin. "A Reporter at Large: American Royal." *New Yorker*, September 26, 1983, p. 57.

Trussell, Robert C. "Barbecue: a taste of ingenuity." *Kansas City Star*, November 1, 1985.

Untersee, Warner J. *Flood Disaster: Kansas City — 1951*. Prologue by Grier Lowery. 2nd edition. Kansas City: Warner Studio, n.d.

The WPA Guide to 1930s Missouri. Compiled by Workers of the Writers' Program of the Work Projects Administration in the State of Missouri. Originally pub. 1941. Lawrence: University Press of Kansas, 1986.

Weatherman, Lynn. "Lady Jane." *Saddle & Bridle*, August 1981.

Weatherman, Lynn. "1931: A Year of Tragedy." *The American Saddlebred*, May/June 1991, p. 75.

"Will Shriver and the American Saddle Horse." Press release, 1976.

Wilson, William H. *The City Beautiful Movement in Kansas City* (2nd edition). Kansas City: Lowell Press, 1990.

Zeeck, David. "Kemper Arena Called Tribute to Midwest." Kansas City Star [?], Friday, October 18, 1974 [?].

Additional Bibliography: Books about Kansas City.

Bradley, Lenore K. *Corinthian Hall: An American Palace on Gladstone*. Kansas City, Missouri: Lowell Press with the Kansas City Museum, 1981.

Brown, A. Theodore and Lyle W. Dorsett. *K. C.: A History of Kansas City, Missouri*. The Western Urban History Series, v. 2. Boulder, Colorado: Pruett, 1978.

Cox, Jack, comp. *Kansas City — The Way We Were*. Kansas City, Missouri: Wilborn and Assocites, c. 1981.

DeAngelo, Dory. *Passages through time: Stories about Kansas City, Missouri and Its Northeast Neigh-*

Bibliography

borhood. Kansas City, Missouri: Tapestry, 1992.

DeAngelo, Dory. *Voices Across Time: Profiles of Kansas City's Early Residents*. Kansas City, Missouri: Tapestry, 1987.

DeAngelo, Dory. *What About Kansas City! A Historical Handbook*. Kansas City, Missouri: Two Lane, 1995.

DeAngelo, Dory and Jane Fifield Flynn. *Kansas City Style: A Social and Cultural History of Kansas City as Seen through its Lost Architecture*. Kansas City, Missouri: Fifield, 1992.

Deatherage, Charles P. *History of Greater Kansas City: Early History 1842 to 1870*. Kansas City, Missouri: Interstate, 1928. [This was intended to be the first of three volumes, but the other books were never published.]

Dorsett, Lyle W. *The Pendergast Machine*. 1968. Lincoln: University of Nebraska Press, 1980.

Ehrlich, George. *Kansas City, Missouri: An Architectural History, 1826-1976*. Kansas City, Missouri: Historic Kansas City Foundation, 1979.

Flynn, Jane Fifield. *Kansas City Women of Independent Minds*. Kansas City, Missouri: Fifield, 1992.

Fowler, Dick. *Leaders in Our Town*. Kansas City: Burd & Fletcher, [1952].

Haskell, Henry C., Jr. and Richard B. Fowler. *City of the Future*. Kansas City: Frank Glenn, 1950.

Hoffhaus, Charles E. *Chez les Canses: Three Centuries at Kawsmouth: The French Foundations of Metropolitan Kansas City*. Kansas City, Missouri: Lowell Press, 1984.

Jones, Linda Newcom. *The Longview We Remember*. n. p.: Storm Ridge, 1990.

Kansas City: A Celebration of the Heartland. Kansas City, Missouri: Hallmark, 1991.

Keeley, Mary Paxton. *Back in Independence*. Chillicothe, Missouri: Community Press, 1992.

Larsen, Lawrence Harold and Nancy J. Hulston. *Pendergast*. Columbia: University of Missouri Press, 1997.

A Legacy of Design: An Historical Survey of the Kansas City, Missouri Parks and Boulevards System, 1893-1940. Ed. Janice Lee, David Boutros, Charlotte R. White and Deon Wolfenbarger. Kansas City, Missouri: Kansas City Center for Design Education and Research, in cooperation with the Western Historical Manuscript Collection-Kansas City, 1995.

McCullough, David. *Truman*. New York: Simon & Schuster, 1992.

Miller, Patricia Cleary. *Westport, Missouri's Port of Many Returns*. Kansas City, Missouri: Lowell Press, 1983.

Milligan, Maurice Morton. *Missouri Waltz: The Inside Story of the Pendergast Machine by the Man Who Smashed It.* New York: Charles Scribner's Sons, 1948.

Mobley, Jane and Nancy Whitnell Harris. *A City Within a Park: One Hundred Years of Parks and Boulevards in Kansas City, Missouri.* Kansas City, Missouri: American Society of Landscape Architects and the Kansas City, Missouri Board of Parks and Recreation Commissioners, 1991.

Monnett, Howard N. *Action before Westport, 1864.* Niwot, Colorado: University Press of Colorado, 1995.

The 1951 Flood in Greater Kansas City: A Picture Review. Kansas City, Missouri: Brown, White, Lowell, 1951.

Ray, Mrs. Sam [Mildred]. *Postcards from Old Kansas City.* Ed. Joan Michalaki. Kansas City, Missouri: Historic Kansas City Foundation, 1980.

Ray, Mrs. Sam [Mildred]. *Postcards from Old Kansas City II.* Ed. Doran L. Cart. Kansas City, Missouri: Historic Kansas City Foundation, 1987.

Rebello, Joseph. "Kansas City calls home the cows . . ." *Kansas City Star Magazine* [On-line], Sunday, October 25, 1992.

Reddig, William M. *Tom's Town: Kansas City and the Pendergast Legend.* 1947. Columbia: University of Missouri Press, 1986.

Serda, Daniel. *A Blow to the Spirit: The Kaw River Flood of 1951 in Perspective.* Kansas City, Missouri: Midwest Research Institute, 1993.

Serda, Daniel. *Boston Investors and the Early Development of Kansas City, Missouri.* Kansas City, Missouri: Midwest Research Institute, 1992.

Simmons, Donald H., ed. *Centennial History of Argentine, Kansas City, Kansas 1880-1980.* Kansas City, Kansas: Simmons Funeral Home, 1980.

Spalding, C. C. *Annals of the City of Kansas.* 1898. Kansas City, Missouri: Frank Glenn, 1950.

Tales of the 1951 Flood. Kansas City, Kansas: Wyandotte County Genealogical Society, 1993.

Thomas, Tracy and Walt Bodine. *Right Here in River City: A Portrait of Kansas City.* Garden City, New York: Doubleday, 1976.

Unger, Robert. *The Union Station Massacre: The Original Sin of J. Edgar Hoover's FBI.* Kansas City, Missouri: Andrews McMeel, 1997.

Westlake, Carrie. *Kansas City Missouri: Its History and Its People* (3 vol.), Chicago: S. J. Clarke, 1908.

Bibliography

Wilborn, Chris, comp. *Where the Streetcar Stops*. Kansas City, Missouri: Wilborn & Associates Photographers, 1991.

Wilson, William H. *The City Beautiful Movement in Kansas City*. 2nd ed. Kansas City, Missouri: Lowell Press, 1990.

Wolferman, Kristie C. *The Nelson-Atkins Museum of Art: Culture Comes to Kansas City*. Columbia: University of Missouri Press, 1993.

Worley, William S. *J. C. Nichols and the Shaping of Kansas City: Innovation in Planned Residential Communities*. Columbia: University of Missouri Press, 1990.

Worley, William S. *The Plaza, First and Always*. Foreword by Miller Nichols. Lenexa, Kansas: Addax, 1997.

Young, William H. and Nathan B. Young. *Your Kansas City and Mine, 1850-1950*. Kansas City, Missouri: Midwest Afro-American Genealogical Society, 1997.